THE NEW CONSOLIDATED ATLAS OF THE NEAR SIDE OF THE MOON

Volume 1
Northern Hemisphere

For Helene and Christopher

The New Consolidated Atlas of the Near Side of the Moon
Volume 1, Northern Hemisphere

Image Credits and Acknowledgements

For the maps, images of NASA LRO/LROC have been used taken from https://quickmap.lroc.asu.edu and https://wms.lroc.asu.edu/lroc/view_rdr_product/. I acknowledge the use of imagery from Lunar QuickMap (https://quickmap.lroc.asu.edu), a collaboration between NASA, Arizona State University & Applied Coherent Technology Corp. I also acknowledge the use of imagery from the LROC website (https://www.lroc.asu.edu/). The website has a wealth of information on the LROC mission and stunning images beyond belief.

The images on pages 6 (map excerpt), 7, 8 and maps 1-38 are derived from the LROC PDS archive
https://wms.lroc.asu.edu/lroc/view_rdr_product/WAC_GLOBAL_O000N0000_100M.

The images of maps Detail 1 – 8 are derived from
https://quickmap.lroc.asu.edu.

The cover image, title page image, images on pages 4, 6 (Plato) and 100 are from the author.

Contact: Consolidatedmoonatlas@outlook.com

ISBN 979-88-66615-88-9

Table of Contents

Introduction

Every so often an observer of the Moon will find him/herself looking for the name of a certain feature that is not noted on one map or another. At some point it becomes apparent that the current, official, naming convention of the Moon leaves many features unnamed. More importantly, it leaves many features unnamed that previously had names. Of course every feature can be identified using coordinates, but in conversations about the Moon it will become quite tedious using coordinates instead of names.

There are many fine atlases of the Moon, each serving their own purpose. Some show drawings of the Moon, giving the map relevance no matter the actual light and shadows of any particular day. Others have based the maps on photographs, showing each detail exactly as seen through the telescope or on photographs (depending on the optical system used).

However, to date I have found no atlas that has as many designated features as possible so that observers can note and share their observations with ease; including the designations of the modern naming convention, completed with the old names for those features that currently lack an official name.

The basis for all names in this atlas are names that have been used by the IAU, occupying itself with the subject since its founding in 1919, and/or NASA, actively involved in the subject, especially during the 60's and 70's, due to the need for accurate lunar maps used for the space program.

Perhaps the best starting point is the publication "Named Lunar Formations" (NLF), by M.A. Blagg and K. Müller from 1935. This was based on the result of an enormous task to list, compare and consolidate the main lunar maps of that time (Neison, Schmidt, Beer and Mädler) and their nomenclature, published in 1913 under the name "Collated List of Lunar Formations" and several additions from e.g. Franz, Saunders and from Wilkins earlier map from 1924[1]. This list was subsequently improved by "The System of Lunar Craters" by D.W.G. Arthur et al. These were published in the Communications of the Lunar and Planetary Laboratory, in four publications each focussing on a separate quadrant from 1963-1966. It notes that the NLF did not achieve wide usage likely due to the facts that it was not widely distributed and the poor legibility of parts of the map. It also notes that the NLF "... *contains craters that do not exist, craters with two designations, craters with identical designations, illogical situations in which objects are associated with named craters which lie beyond other named craters, and other similar defects*"[2]. The document goes on to explain that many of the Greek elevations from NLF were discontinued because it followed Mädler and Schmidt in the practise to also use Greek letters for parts of the walls of craters. The contents of The System of Lunar Craters was approved by the IAU in 1964 and 1967[3].

---

[1] Mary A. Blagg, K. Müller, "Named Lunar Formations"

[2] D.W.G. Arthur et al, "No. 30 The System of Lunar Craters, Quadrant 1"

[3] Leif E. Andersson, Ewen A. Whitaker, "NASA Catalogue of Lunar Nomenclature" (RP-1097)

In 1982 the NASA Catalogue of Lunar Nomenclature (Reference Publication 1097) incorporates all changes made to the list of approved names of the IAU up to mid 1981. It states "This catalogue may therefore be used with full confidence by researchers, cartographers, etc. Its listings supersede all earlier catalogues and maps; any names found on these older documents that do not appear here have been dropped or substituted for good reason, and should not be used" [4].

The document does not include any Greek letter elevations, nor Roman numeral designations of rilles. It does feature all letter designation subsidiary craters, which were dropped by the IAU between 1973 and 1976. In 1973 the IAU also dropped the Greek lettered elevations and Roman numbered rilles, but these were not reinstated in 1976.

That status quo sets the stage for the current list of named formations on the Moon. It also explains why we are faced with a situation where hundreds of features are currently unnamed, that used to have an official name in the mid 20th century. Current maps or other publications largely stick to the official IAU list, but one can regularly find references to discontinued names, where users have found it necessary. This shows the need for such features to be identifiable and eventually officially named.

Therefore this atlas also shows the names of features that are currently unnamed in the official IAU list, but which have had official names in the past, allowing lunar observers greater ease of exchanging information. From the different historical sources that are at hand, the most optimal sources for the extended nomenclature usage have been the Atlas and Gazetteer of the Near Side of the Moon and the AIC/LAC maps from NASA. These two (sets of) documents were produced in the "sweet spot" of Lunar nomenclature; after the publications of "The System of Lunar Craters" and before the revision of Lunar nomenclature done by the IAU in 1973.

In this atlas the reader will find both official and unofficial names:

- Names and numbers in white, blue or red are the official IAU names as found on their LAC maps.
- Names and numbers in yellow are names from older / other sources, most notably:
    - Names of elevations using the Greek alphabet.
    - Names of Craters with two roman letters, e.g. "KA" as a sub crater of "K"
    - Individually numbered Rimae (I/II/III). Please note that these are sometimes shown as only roman numerals.

The maps show main features written out and sub features that are numbered, all names appear on the adjacent page. In principle a number is either on/inside the feature or directly right of the feature. If it would be unclear which feature the number refers to a line has been drawn to the feature.

Users of Rükl's Atlas of the Moon, will notice the compatibility of the maps between both. I have chosen to make the maps largely compatible to allow visual observers to match their notes.

This document is solely an atlas, no further background information has been given. For further background reading and/or research I can recommend the following books:

- Anyone who would like to read more about the background of lunar features is very well served by reading Luna Cognita by Robert A. Garfinkle. This set of three books is a vast and comprehensive compendium about the Moon with hundreds of pages devoted to the description of every imaginable detail of the Moon.
- The book "Mapping and Naming the Moon" by Ewen A. Whitaker is a great resource for anyone interested in the history and evolution of lunar maps and naming. Mr. Whitaker is also a co-author of an atlas frequently used in the creation of this book; The Atlas and Gazetteer of the Near Side of the Moon
- John Moore has written several books on craters and other features of the Moon; Craters of the Near Side Moon and Features of the Near Side Moon are two books I can recommend for anyone needing background information on lunar features.

---

[4] Leif E. Andersson, Ewen A. Whitaker, "NASA Catalogue of Lunar Nomenclature" (RP-1097)

Features and their appearance in this atlas:

The image on the previous page shows an excerpt of map 22.

Official main features	"Conon"	Crater on the top right of the image
Official sub features	"12"	Small craterlet above Conon
Official sub features	"23"	This sub feature has a dotted line around to show the approximate area
Official Sinus/Mare, etc	"Sinus Fidei"	Sinus in the bottom right quadrant
Official mountain range	"Montes Apenninus"	Mountain range stretching across image
Unofficial sub feature	"45"	Crater left of Conon, in yellow

A magenta square is displayed on certain maps, referring to a detail map at the end of the book. Some of these maps also show landing sites of Apollo missions, here a red triangle shows the location where the lunar module landed.

Numbering is done from top to bottom, first numbering the official features and secondly the unofficial features.

1
2
3
4
8
9
10
11
12
17
18
19
20
21
22
28
29
30
31
32
33

4 5 6 7

12 13 14 15 16

22 23 24 25 26 27

33 34 35 36 37 38

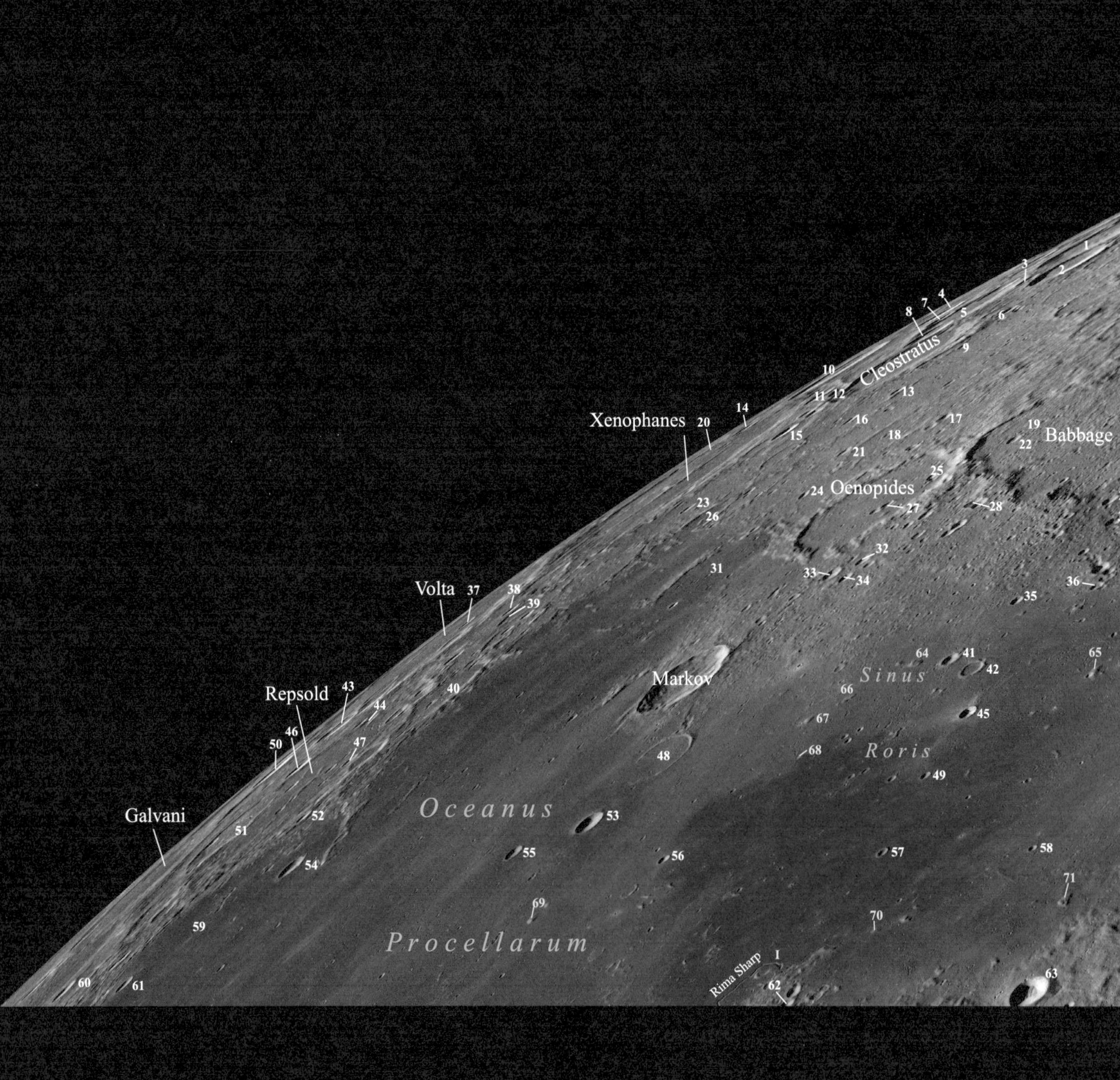

1
2
3
4
7
5
8
6
Cleostratus
9
10
11
12
13
Xenophanes
20
14
15
16
17
18
19
22
Babbage
21
25
24
Oenopides
27
28
29
30
23
26
31
32
33
34
36
35
Volta
37
38
39
Markov
64
41
42
65
Sinus
40
Repsold
43
44
66
45
67
46
50
47
48
68
Roris
49
Galvani
51
52
Oceanus
53
54
55
56
57
58
71
69
70
59
Procellarum
Rima Sharp
1
60
61
62
63

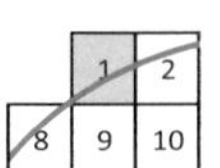

1

No.	Name	No.	Name	Name	No.	Name	No.
	Cleostratus	52	Repsold V	Babbage		Sharp ξ	71
	Harpalus	53	Markov E	Babbage B	28	Sinus Roris	
	Babbage	54	Repsold R	Babbage D	19	South E	30
	Oenopides	55	Markov F	Babbage E	22	South F	29
	Volta	56	Markov G	Boole A	1	South G	35
	Markov	57	Harpalus T	Cleostratus		South M	36
	Sinus Roris	58	Sharp W	Cleostratus A	2	Volta	
	Repsold	59	Repsold C	Cleostratus E	8	Volta B	37
	Galvani	60	Repsold S	Cleostratus F	5	Volta D	43
	Oceanus Procellarum	61	Repsold T	Cleostratus H	7	Xenophanes	
	Rima Sharp	62	Sharp U	Cleostratus J	4	Xenophanes A	10
1	Boole A	63	Sharp A	Cleostratus L	3	Xenophanes B	11
2	Cleostratus A	64	Markov λ	Cleostratus M	6	Xenophanes C	12
3	Cleostratus L	65	Harpalus γ	Cleostratus N	9	Xenophanes D	15
4	Cleostratus J	66	Markov μ	Cleostratus P	13	Xenophanes E	20
5	Cleostratus F	67	Markov θ	Cleostratus R	16	Xenophanes F	26
6	Cleostratus M	68	Markov τ	Galvani		Xenophanes G	23
7	Cleostratus H	69	Markov σ	Harpalus E	45	Xenophanes K	14
8	Cleostratus E	70	Sharp η	Harpalus G	42	Xenophanes L	39
9	Cleostratus N	71	Sharp ξ	Harpalus H	41	Xenophanes M	38
10	Xenophanes A			Harpalus S	49		
11	Xenophanes B			Harpalus T	57		
12	Xenophanes C			Harpalus γ	65		
13	Cleostratus P			Langley J	50		
14	Xenophanes K			Markov			
15	Xenophanes D			Markov E	53		
16	Cleostratus R			Markov F	55		
17	Oenopides Z			Markov G	56		
18	Oenopides B			Markov U	48		
19	Babbage D			Markov θ	67		
20	Xenophanes E			Markov λ	64		
21	Oenopides S			Markov μ	66		
22	Babbage E			Markov σ	69		
23	Xenophanes G			Markov τ	68		
24	Oenopides T			Oceanus Procellarum			
25	Oenopides X			Oenopides			
26	Xenophanes F			Oenopides B	18		
27	Oenopides Y			Oenopides K	32		
28	Babbage B			Oenopides L	33		
29	South F			Oenopides M	34		
30	South E			Oenopides R	31		
31	Oenopides R			Oenopides S	21		
32	Oenopides K			Oenopides T	24		
33	Oenopides L			Oenopides X	25		
34	Oenopides M			Oenopides Y	27		
35	South G			Oenopides Z	17		
36	South M			Repsold			
37	Volta B			Repsold A	47		
38	Xenophanes M			Repsold B	40		
39	Xenophanes L			Repsold C	59		
40	Repsold B			Repsold G	51		
41	Harpalus H			Repsold H	46		
42	Harpalus G			Repsold R	54		
43	Volta D			Repsold S	60		
44	Repsold W			Repsold T	61		
45	Harpalus E			Repsold V	52		
46	Repsold H			Repsold W	44		
47	Repsold A			Rima Sharp			
48	Markov U			Sharp A	63		
49	Harpalus S			Sharp U	62		
50	Langley J			Sharp W	58		
51	Repsold G			Sharp η	70		

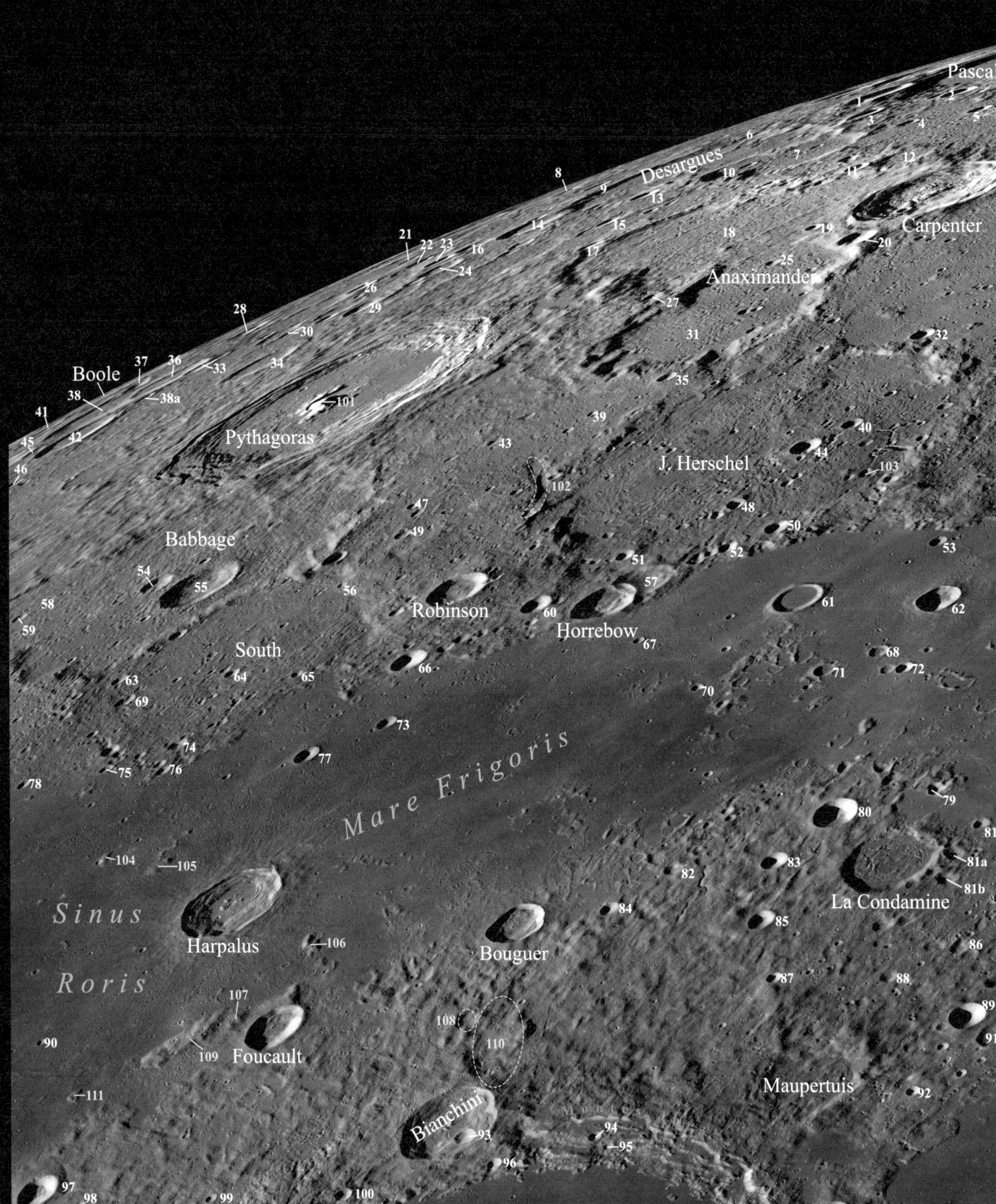

Pascal
1
2
3
4
5
6
7
Desargues
10
11
12
8
9
13
14
15
18
19
20
Carpenter
21
22
23
16
17
24
25
Anaximander
26
27
29
28
30
31
32
37
36
33
34
Boole
35
38
38a
101
41
42
39
40
Pythagoras
43
45
44
J. Herschel
46
102
103
47
48
50
49
Babbage
53
51
52
54
57
55
56
Robinson
60
61
62
58
59
Horrebow
67
South
68
72
63
64
65
66
71
70
69
73
Mare Frigoris
74
77
75
76
78
79
80
81
104
105
81a
83
82
81b
Sinus
La Condamine
84
85
Harpalus
106
86
Bouguer
Roris
87
88
107
108
89
110
90
109
Foucault
91
Maupertuis
111
92
Bianchini
94
93
95
96
97
98
99
100

1	2	3
9	10	11

2

No.	Name	No.	Name	Name	No.	Name	No.
	Pascal	47	Babbage U	Anaximander		J. Herschel	
	Desargues	48	J. Herschel L	Anaximander A	20	J. Herschel B	52
	Carpenter	49	Babbage X	Anaximander B	18	J. Herschel C	44
	Anaximander	50	J. Herschel D	Anaximander D	31	J. Herschel D	50
	Boole	51	Horrebow G	Anaximander H	32	J. Herschel F	61
	Pythagoras	52	J. Herschel B	Anaximander R	27	J. Herschel G	103
	J. Herschel	53	J. Herschel N	Anaximander S	19	J. Herschel K	40
	Babbage	54	Babbage C	Anaximander T	25	J. Herschel L	48
	Robinson	55	Babbage A	Anaximander U	35	J. Herschel M	71
	Horrebow	56	South K	Babbage		J. Herschel N	53
	South	57	Horrebow A	Babbage A	55	La Condamine	
	La Condamine	58	Babbage D	Babbage C	54	La Condamine A	80
	Harpalus	59	Babbage E	Babbage D	58	La Condamine B	62
	Mare Frigoris	60	Horrebow B	Babbage E	59	La Condamine C	85
	Sinus Roris	61	J. Herschel F	Babbage U	47	La Condamine D	83
	Bouguer	62	La Condamine B	Babbage X	49	La Condamine E	68
	Foucault	63	South F	Bianchini		La Condamine F	72
	Maupertuis	64	South A	Bianchini A	108	La Condamine G	79
	Bianchini	65	South H	Bianchini D	100	La Condamine H	81b
	Montes Jura	66	South B	Bianchini H	96	La Condamine K	86
1	Pascal A	67	Horrebow D	Bianchini M	95	La Condamine L	81a
2	Pascal G	68	La Condamine E	Bianchini N	94	La Condamine M	81
3	Pascal J	69	South E	Bianchini P	110	Mare Frigoris	
4	Carpenter Y	70	Horrebow C	Bianchini W	93	Maupertuis	
5	Carpenter W	71	J. Herschel M	Boole		Maupertuis A	89
6	Desargues A	72	La Condamine F	Boole A	38	Maupertuis B	88
7	Desargues B	73	Harpalus B	Boole B	38a	Maupertuis C	91
8	Desargues L	74	South C	Boole C	28	Maupertuis K	92
9	Desargues C	75	South M	Boole D	37	Maupertuis L	87
10	Desargues E	76	South D	Boole E	41	Montes Jura	
11	Carpenter T	77	Harpalus C	Boole F	36	Pascal	
12	Carpenter U	78	South G	Boole R	33	Pascal A	1
13	Desargues D	79	La Condamine G	Bouguer		Pascal G	2
14	Desargues M	80	La Condamine A	Bouguer A	84	Pascal J	3
15	Desargues K	81	La Condamine M	Bouguer B	82	Pythagoras	
16	Pythagoras G	81a	La Condamine L	Carpenter		Pythagoras B	29
17	Pythagoras S	81b	La Condamine H	Carpenter T	11	Pythagoras D	34
18	Anaximander B	82	Bouguer B	Carpenter U	12	Pythagoras E	102
19	Anaximander S	83	La Condamine D	Carpenter W	5	Pythagoras G	16
20	Anaximander A	84	Bouguer A	Carpenter Y	4	Pythagoras H	24
21	Pythagoras M	85	La Condamine C	Cleostratus A	42	Pythagoras K	23
22	Pythagoras L	86	La Condamine K	Cleostratus L	45	Pythagoras L	22
23	Pythagoras K	87	Maupertuis L	Cleostratus M	46	Pythagoras M	21
24	Pythagoras H	88	Maupertuis B	Desargues		Pythagoras N	26
25	Anaximander T	89	Maupertuis A	Desargues A	6	Pythagoras P	30
26	Pythagoras N	90	Sharp W	Desargues B	7	Pythagoras S	17
27	Anaximander R	91	Maupertuis C	Desargues C	9	Pythagoras T	43
28	Boole C	92	Maupertuis K	Desargues D	13	Pythagoras W	39
29	Pythagoras B	93	Bianchini W	Desargues E	10	Pythagoras α	101
30	Pythagoras P	94	Bianchini N	Desargues K	15	Robinson	
31	Anaximander D	95	Bianchini M	Desargues L	8	Sharp A	97
32	Anaximander H	96	Bianchini H	Desargues M	14	Sharp K	99
33	Boole R	97	Sharp A	Foucault		Sharp M	98
34	Pythagoras D	98	Sharp M	Foucault β	109	Sharp W	90
35	Anaximander U	99	Sharp K	Foucault γ	107	Sharp ξ	111
36	Boole F	100	Bianchini D	Harpalus		Sinus Roris	
37	Boole D	101	Pythagoras α	Harpalus B	73	South	
38	Boole A	102	Pythagoras E	Harpalus C	77	South A	64
38a	Boole B	103	J. Herschel G	Harpalus γ	104	South B	66
39	Pythagoras W	104	Harpalus γ	Harpalus λ	106	South C	74
40	J. Herschel K	105	Harpalus ξ	Harpalus ξ	105	South D	76
41	Boole E	106	Harpalus λ	Horrebow		South E	69
42	Cleostratus A	107	Foucault γ	Horrebow A	57	South F	63
43	Pythagoras T	108	Bianchini A	Horrebow B	60	South G	78
44	J. Herschel C	109	Foucault β	Horrebow C	70	South H	65
45	Cleostratus L	110	Bianchini P	Horrebow D	67	South K	56
46	Cleostratus M	111	Sharp ξ	Horrebow G	51	South M	75

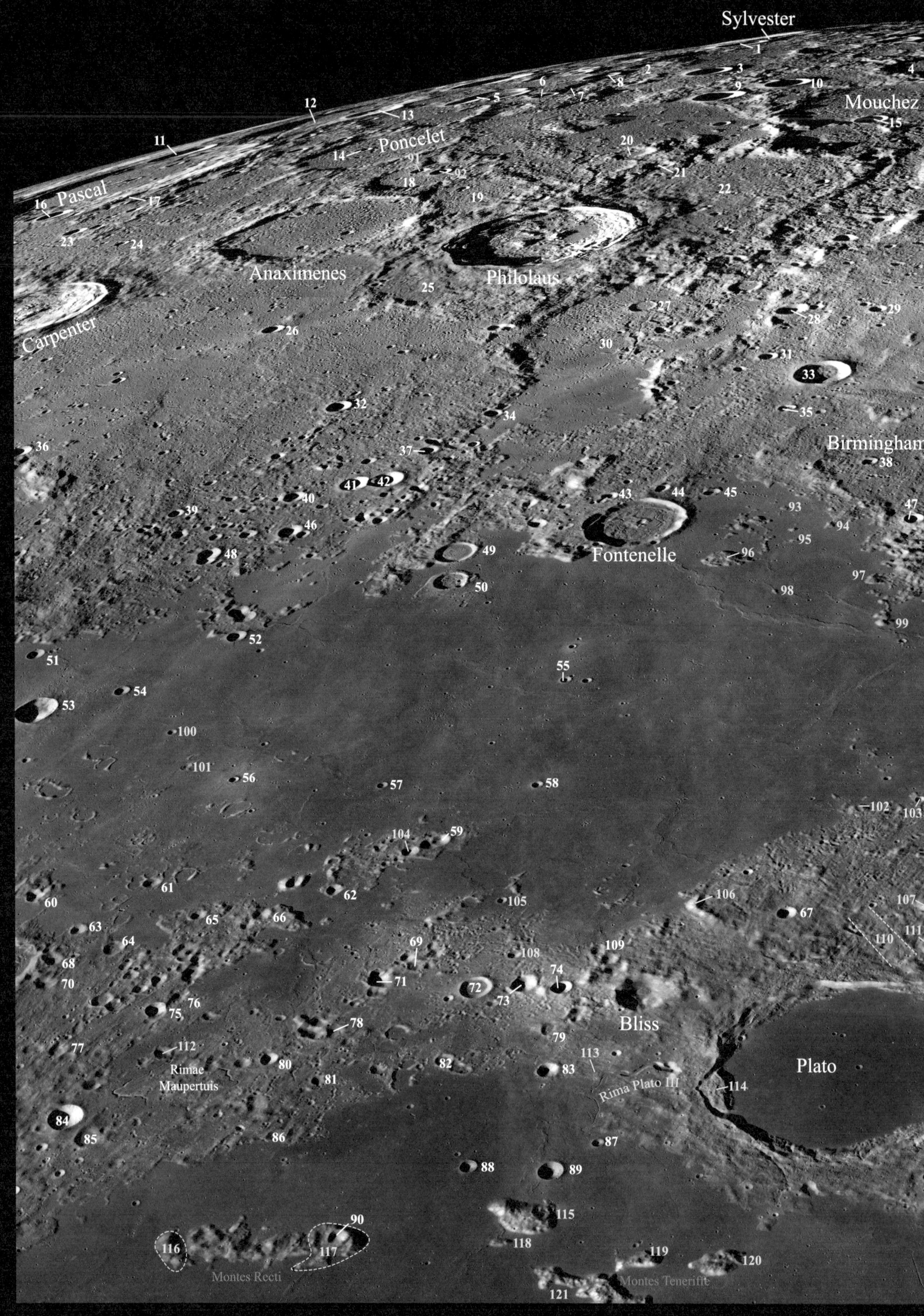

Sylvester
1
2
3
4
5
6
7
8
9
10
Mouchez
15
11
12
13
Poncelet
14
91
92
18
19
20
21
22
Pascal
16
17
23
24
Anaximenes
Philolaus
25
Carpenter
26
27
28
29
30
31
32
33
34
35
36
37
Birmingham
38
39
40
41
42
43
44
45
46
47
93
94
95
48
49
Fontenelle
96
97
50
98
99
51
52
53
54
55
100
101
56
57
58
102
103
104
59
60
61
62
105
106
107
67
63
64
65
66
69
108
109
110
111
68
70
71
72
73
74
75
76
Bliss
77
78
79
112
Rimae
Maupertuis
80
81
82
83
113
Rima Plato III
114
Plato
84
85
86
87
88
89
115
90
116
117
118
119
120
Montes Recti
Montes Teneriffe
121

2	3	4
10	11	12

3

No.	Name	No.	Name	Name	No.	Name	No.
	Sylvester	54	La Condamine T	Anaximander H	36	Maupertuis C	[illegible]
	Poncelet	55	Fontenelle G	Anaximenes		Montes Recti	
	[illegible]	56	La Condamine S	Anaximenes B	26	Montes Recti B	90
	Pascal	57	La Condamine X	Anaximenes E	32	Montes Recti β	117
	Carpenter	58	Plato W	Anaximenes G	19	Montes Recti ϵ	116
	Anaximenes	59	La Condamine J	Anaximenes H	18	Montes Teneriffe	
	Philolaus	60	La Condamine G	Anaximenes HA	92	Montes Teneriffe α	119
	Birmingham	61	La Condamine O	Anaximenes HB	91	Montes Teneriffe δ	121
	Fontenelle	62	La Condamine R	Birmingham		Montes Teneriffe ϵ	115
	Bliss	63	La Condamine M	Birmingham B	47	Montes Teneriffe ι	120
	Plato	64	La Condamine N	Birmingham K	38	Montes Teneriffe ω	118
	Rimae Maupertuis	65	La Condamine V	Birmingham β	97	Mouchez	
	Montes Recti	66	La Condamine U	Birmingham ϵ	96	Mouchez A	4
	Montes Teneriffe	67	Plato T	Birmingham ζ	94	Mouchez C	15
	Rima Plato III	68	La Condamine L	Birmingham η	93	Mouchez J	10
1	Sylvester N	69	Plato R	Birmingham λ	99	Mouchez L	9
2	Poncelet P	70	La Condamine H	Birmingham μ	98	Mouchez M	3
3	Mouchez M	71	Plato C	Birmingham ξ	95	Pascal	
4	Mouchez A	72	Plato B	Bliss		Pascal F	11
5	Poncelet B	73	Plato Y	Carpenter		Pascal G	16
6	Poncelet S	74	Plato M	Carpenter V	24	Pascal L	17
7	Poncelet R	75	La Condamine Q	Carpenter W	23	Philolaus	
8	Poncelet Q	76	La Condamine P	Fontenelle		Philolaus B	27
9	Mouchez L	77	La Condamine K	Fontenelle A	33	Philolaus C	25
10	Mouchez J	78	Laplace M	Fontenelle B	50	Philolaus D	22
11	Pascal F	79	Plato O	Fontenelle C	42	Philolaus E	28
12	Poncelet C	80	Laplace L	Fontenelle D	49	Philolaus F	31
13	Poncelet D	81	Laplace B	Fontenelle F	41	Philolaus G	30
14	Poncelet H	82	Plato F	Fontenelle G	55	Philolaus U	21
15	Mouchez C	83	Plato P	Fontenelle H	43	Philolaus W	20
16	Pascal G	84	Maupertuis A	Fontenelle K	29	Plato	
17	Pascal L	85	Maupertuis C	Fontenelle L	35	Plato B	72
18	Anaximenes H	86	Laplace E	Fontenelle M	46	Plato BA	108
19	Anaximenes G	87	Plato X	Fontenelle N	40	Plato BB	105
20	Philolaus W	88	Plato E	Fontenelle P	45	Plato C	71
21	Philolaus U	89	Plato D	Fontenelle R	44	Plato D	89
22	Philolaus D	90	Montes Recti B	Fontenelle S	37	Plato E	88
23	Carpenter W	91	Anaximenes HB	Fontenelle T	34	Plato F	82
24	Carpenter V	92	Anaximenes HA	Fontenelle X	52	Plato M	74
25	Philolaus C	93	Birmingham η	J. Herschel N	51	Plato O	79
26	Anaximenes B	94	Birmingham ζ	J. Herschel P	39	Plato P	83
27	Philolaus B	95	Birmingham ξ	J. Herschel R	48	Plato R	69
28	Philolaus E	96	Birmingham ϵ	La Condamine B	53	Plato S	109
29	Fontenelle K	97	Birmingham β	La Condamine G	60	Plato T	67
30	Philolaus G	98	Birmingham μ	La Condamine H	70	Plato VA	103
31	Philolaus F	99	Birmingham λ	La Condamine J	59	Plato W	58
32	Anaximenes E	100	La Condamine TA	La Condamine JA	104	Plato X	87
33	Fontenelle A	101	La Condamine SA	La Condamine K	77	Plato Y	73
34	Fontenelle T	102	Plato ϕ	La Condamine L	68	Plato Z	110
35	Fontenelle L	103	Plato VA	La Condamine M	63	Plato ζ	114
36	Anaximander H	104	La Condamine JA	La Condamine N	64	Plato η	113
37	Fontenelle S	105	Plato BB	La Condamine O	61	Plato ν	107
38	Birmingham K	106	Plato ρ	La Condamine P	76	Plato ρ	106
39	J. Herschel P	107	Plato ν	La Condamine Q	75	Plato σ	111
40	Fontenelle N	108	Plato BA	La Condamine R	62	Plato ϕ	102
41	Fontenelle F	109	Plato S	La Condamine S	56	Poncelet	
42	Fontenelle C	110	Plato Z	La Condamine SA	101	Poncelet B	5
43	Fontenelle H	111	Plato σ	La Condamine T	54	Poncelet C	12
44	Fontenelle R	112	Laplace HA	La Condamine TA	100	Poncelet D	13
45	Fontenelle P	113	Plato η	La Condamine U	66	Poncelet H	14
46	Fontenelle M	114	Plato ζ	La Condamine V	65	Poncelet P	2
47	Birmingham B	115	Montes Teneriffe ϵ	La Condamine X	57	Poncelet Q	8
48	J. Herschel R	116	Montes Recti ϵ	Laplace B	81	Poncelet R	7
49	Fontenelle D	117	Montes Recti β	Laplace E	86	Poncelet S	6
50	Fontenelle B	118	Montes Teneriffe ω	Laplace HA	112	Rima Plato III	
51	J. Herschel N	119	Montes Teneriffe α	Laplace L	80	Rimae Maupertuis	
52	Fontenelle X	120	Montes Teneriffe ι	Laplace M	78	Sylvester	
53	La Condamine B	121	Montes Teneriffe δ	Maupertuis A	84	Sylvester N	1

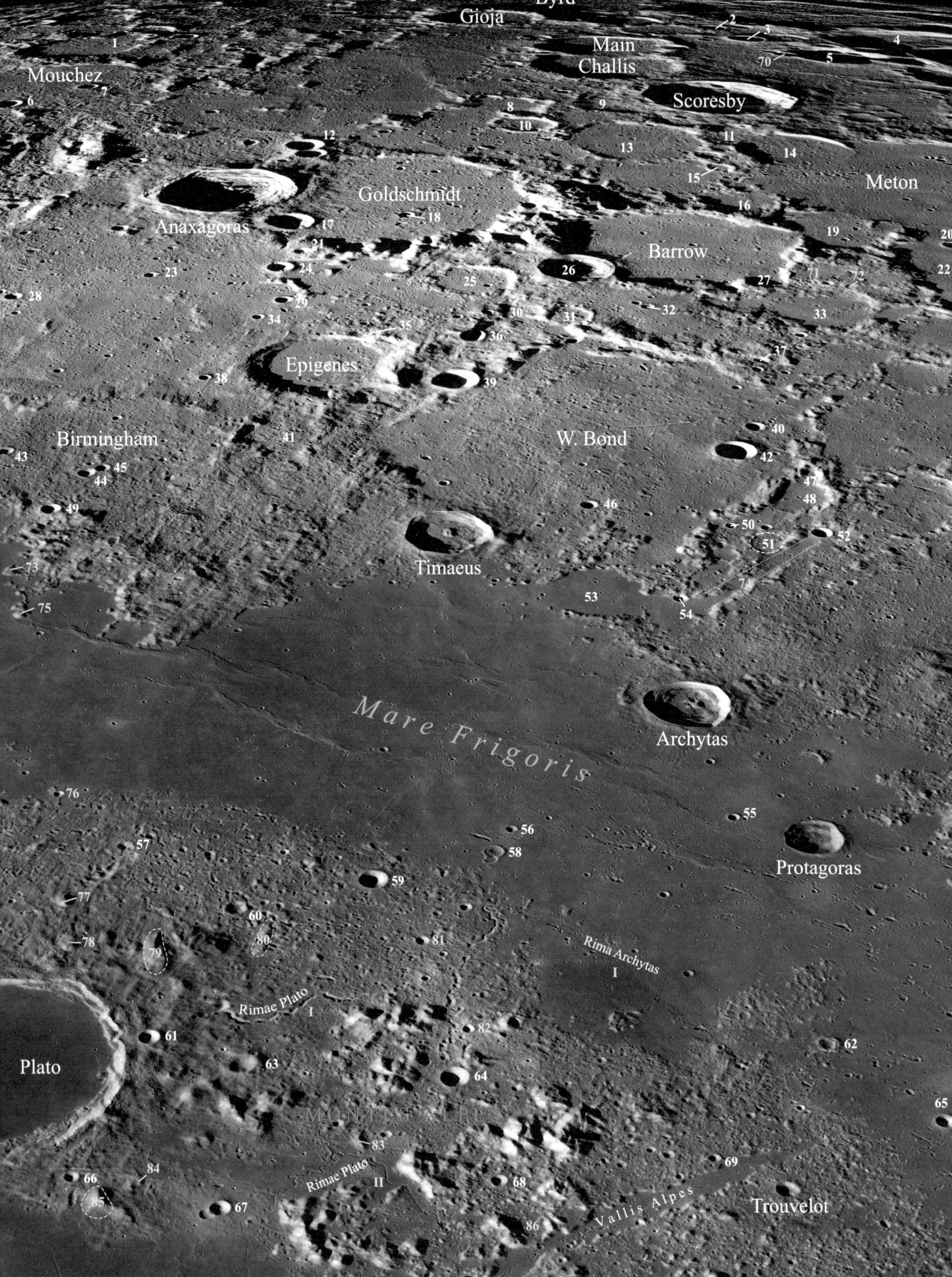

Byrd
Gioja
2
3
4
1
Main
Challis
70
5
Mouchez
7
6
8
9
Scoresby
10
12
11
13
14
15
Meton
Goldschmidt
16
Anaxagoras
17
18
19
20
21
Barrow
24
26
22
23
25
27
71
72
28
29
30
31
32
33
34
35
36
37
Epigenes
38
39
40
Birmingham
41
W. Bond
42
43
45
44
47
48
49
46
50
51
52
Timaeus
73
74
53
75
54
Mare Frigoris
Archytas
76
55
56
57
58
Protagoras
59
77
60
80
81
78
79
Rima Archytas
I
Rimae Plato
I
61
82
62
Plato
63
64
65
MONTES ALPES
83
84
69
66
68
Rimae Plato
II
85
67
Vallis Alpes
Trouvelot
86

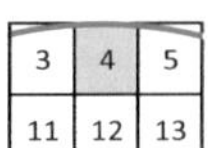

4

No.	Name	No.	Name	Name	No.	Name	No.
	[illegible]	[illegible]	[illegible]	[illegible]	[illegible]	[illegible]	[illegible]
	[illegible]	[illegible]	[illegible]	[illegible]	[illegible]	[illegible]	
	Main	39	Epigenes A	Anaxagoras		Mouchez	
	Challis	40	W. Bond C	Anaxagoras A	17	Mouchez A	1
	Mouchez	41	Epigenes P	Anaxagoras B	23	Mouchez B	7
	Scoresby	42	W. Bond B	Archytas		Mouchez C	6
	Goldschmidt	43	Birmingham K	Archytas B	53	Plato	
	Meton	44	Birmingham H	Archytas C	74	Plato G	61
	Anaxagoras	45	Birmingham G	Archytas G	58	Plato H	59
	Barrow	46	W. Bond D	Archytas K	51	Plato HA	81
	Epigenes	47	W. Bond F	Archytas L	56	Plato J	67
	Birmingham	48	W. Bond E	Archytas U	52	Plato L	63
	W. Bond	49	Birmingham B	Archytas W	54	Plato Q	60
	Timaeus	50	W. Bond G	Barrow		Plato U	66
	Archytas	51	Archytas K	Barrow A	26	Plato V	57
	Mare Frigoris	52	Archytas U	Barrow B	27	Plato VA	76
	Protagoras	53	Archytas B	Barrow C	16	Plato κ	85
	Rima Archytas	54	Archytas W	Barrow E	31	Plato ν	79
	Rima Archytas I	55	Protagoras B	Barrow F	30	Plato π	80
	Rimae Plato	56	Archytas L	Barrow G	25	Plato σ	78
	Rima Plato I	57	Plato V	Barrow H	32	Plato υ	77
	Rima Plato II	58	Archytas G	Barrow K	33	Plato χ	83
	Plato	59	Plato H	Barrow KA	72	Plato ω	84
	Montes Alpes	60	Plato Q	Barrow KB	71	Protagoras	
	Vallis Alpes	61	Plato G	Barrow M	37	Protagoras B	55
	Trouvelot	62	Egede G	Birmingham		Protagoras E	68
1	Mouchez A	63	Plato L	Birmingham B	49	Rima Archytas	
2	Main N	64	Alpes A	Birmingham G	45	Rima Archytas I	
3	Main L	65	Egede B	Birmingham H	44	Rima Plato I	
4	De Sitter M	66	Plato U	Birmingham K	43	Rima Plato II	
5	De Sitter A	67	Plato J	Birmingham β	73	Rimae Plato	
6	Mouchez C	68	Protagoras F	Birmingham λ	75	Scoresby	
7	Mouchez B	69	Trouvelot H	Byrd		Scoresby AA	70
8	Challis A	70	Scoresby AA	Challis		Scoresby K	10
9	Scoresby Q	71	Barrow KB	Challis A	8	Scoresby M	13
10	Scoresby K	72	Barrow KA	De Sitter A	5	Scoresby P	11
11	Scoresby P	73	Birmingham β	De Sitter M	4	Scoresby Q	9
12	Goldschmidt D	74	Archytas C	Egede B	65	Scoresby W	15
13	Scoresby M	75	Birmingham λ	Egede G	62	Timaeus	
14	Meton E	76	Plato VA	Epigenes		Trouvelot	
15	Scoresby W	77	Plato υ	Epigenes A	39	Trouvelot D	86
16	Barrow C	78	Plato σ	Epigenes B	35	Trouvelot H	69
17	Anaxagoras A	79	Plato ν	Epigenes D	36	Vallis Alpes	
18	Goldschmidt A	80	Plato π	Epigenes F	38	W. Bond	
19	Meton F	81	Plato HA	Epigenes G	34	W. Bond B	42
20	Meton B	82	Alpes AB	Epigenes H	29	W. Bond C	40
21	Goldschmidt C	83	Plato χ	Epigenes P	41	W. Bond D	46
22	Meton C	84	Plato ω	Fontenelle K	28	W. Bond E	48
23	Anaxagoras B	85	Plato κ	Gioja		W. Bond F	47
24	Goldschmidt B	86	Trouvelot D	Goldschmidt		W. Bond G	50
25	Barrow G			Goldschmidt A	18		
26	Barrow A			Goldschmidt B	24		
27	Barrow B			Goldschmidt C	21		
28	Fontenelle K			Goldschmidt D	12		
29	Epigenes H			Main			
30	Barrow F			Main L	3		
31	Barrow E			Main N	2		
32	Barrow H			Mare Frigoris			
33	Barrow K			Meton			
34	Epigenes G			Meton B	20		
35	Epigenes B			Meton C	22		
36	Epigenes D			Meton E	14		

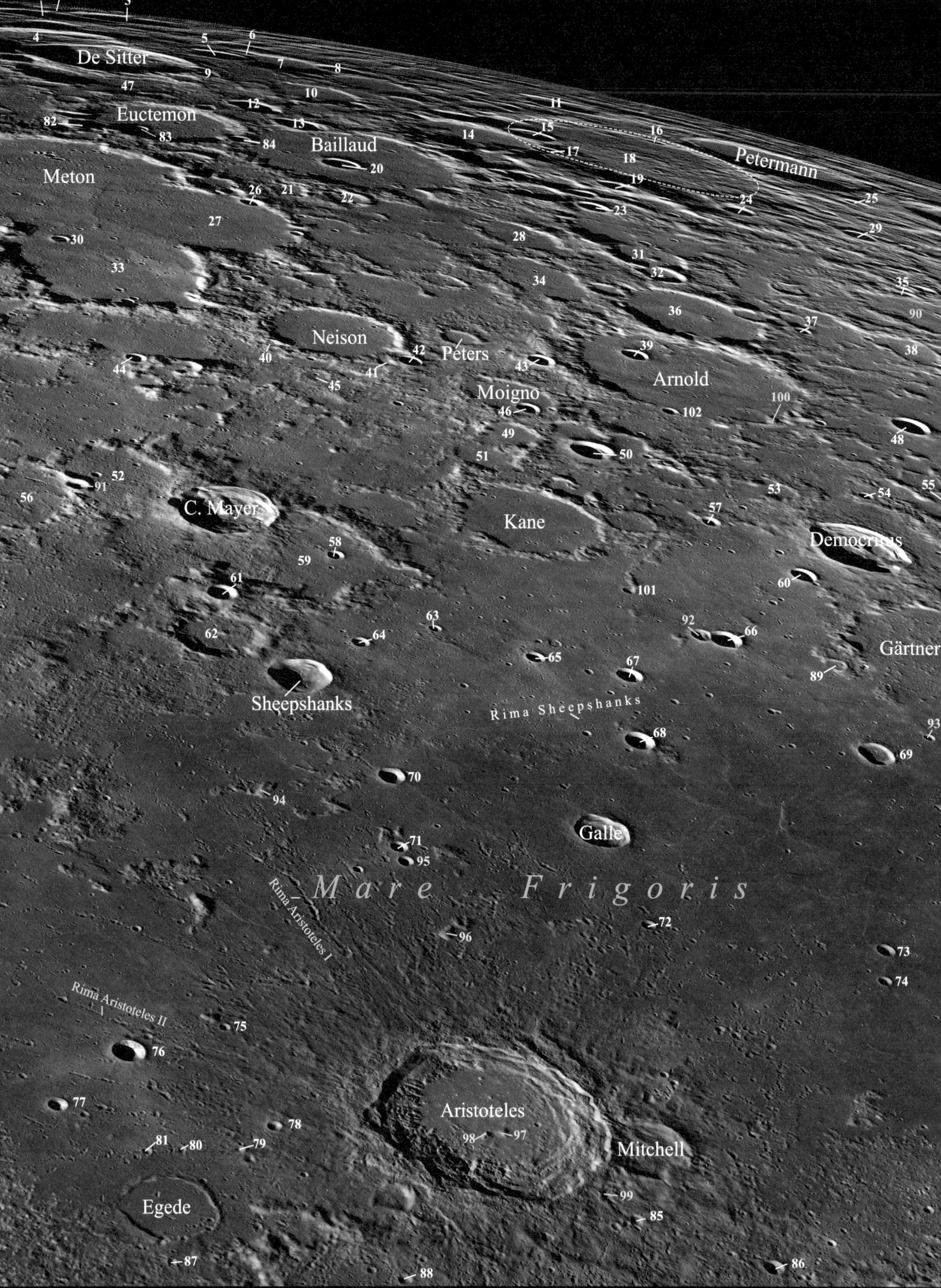

De Sitter
Euctemon
Baillaud
Meton
Petermann
Neison
Peters
Moigno
Arnold
C. Mayer
Kane
Democritus
Gärtner
Sheepshanks
Rima Sheepshanks
Galle
Mare Frigoris
Rima Aristoteles I
Rima Aristoteles II
Aristoteles
Mitchell
Egede

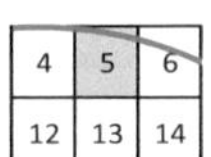

5

No.	Name	No.	Name	Name	No.	Name	No.
	De Sitter	41	Neison B	Archytas D	56	Euctemon D	12
	Euctemon	42	Neison A	Archytas DA	91	Euctemon H	82
	Baillaud	43	Arnold G	Aristoteles		Euctemon K	83
	Petermann	44	Meton W	Aristoteles D	88	Euctemon N	84
	Meton	45	Neison C	Aristoteles M	73	Galle	
	Neison	46	Moigno C	Aristoteles N	74	Galle A	72
	Peters	47	De Sitter L	Aristoteles γ	96	Galle B	71
	Moigno	48	Schwabe G	Aristoteles ζ	98	Galle BA	95
	Arnold	49	Moigno D	Aristoteles θ	97	Galle C	68
	C. Mayer	50	Moigno A	Arnold		Gärtner	
	Kane	51	Moigno B	Arnold A	36	Gärtner C	89
	Democritus	52	C. Mayer H	Arnold E	28	Gärtner F	69
	Sheepshanks	53	Democritus N	Arnold F	39	Gärtner FA	93
	Rima Sheepshanks	54	Democritus M	Arnold G	43	Kane	
	Gärtner	55	Democritus L	Arnold H	23	Kane A	101
	Galle	56	Archytas D	Arnold J	102	Kane F	65
	Mare Frigoris	57	Democritus D	Arnold K	31	Kane G	67
	Rima Aristoteles I	58	C. Mayer F	Arnold L	34	Mare Frigoris	
	Rima Aristoteles II	59	C. Mayer D	Arnold M	37	Meton	
	Aristoteles	60	Democritus A	Arnold N	32	Meton A	21
	Mitchell	61	C. Mayer E	Arnold κ	100	Meton B	30
	Egede	62	C. Mayer B	Baillaud		Meton C	33
1	Nansen D	63	Sheepshanks B	Baillaud A	14	Meton D	27
2	Nansen C	64	Sheepshanks A	Baillaud B	22	Meton G	26
3	Nansen A	65	Kane F	Baillaud C	17	Meton W	44
4	De Sitter M	66	Democritus B	Baillaud D	19	Mitchell	
5	De Sitter F	67	Kane G	Baillaud E	20	Mitchell A	99
6	De Sitter X	68	Galle C	Baillaud F	15	Mitchell B	85
7	De Sitter W	69	Gärtner F	C. Mayer		Mitchell E	86
8	De Sitter V	70	Sheepshanks C	C. Mayer B	62	Moigno	
9	De Sitter G	71	Galle B	C. Mayer D	59	Moigno A	50
10	De Sitter U	72	Galle A	C. Mayer E	61	Moigno B	51
11	Petermann D	73	Aristoteles M	C. Mayer F	58	Moigno C	46
12	Euctemon D	74	Aristoteles N	C. Mayer H	52	Moigno D	49
13	Euctemon C	75	Egede F	De Sitter		Nansen A	3
14	Baillaud A	76	Egede A	De Sitter F	5	Nansen C	2
15	Baillaud F	77	Egede B	De Sitter G	9	Nansen D	1
16	Petermann S	78	Egede C	De Sitter L	47	Neison	
17	Baillaud C	79	Egede M	De Sitter M	4	Neison A	42
18	Petermann R	80	Egede N	De Sitter U	10	Neison B	41
19	Baillaud D	81	Egede E	De Sitter V	8	Neison C	45
20	Baillaud E	82	Euctemon H	De Sitter W	7	Neison D	40
21	Meton A	83	Euctemon K	De Sitter X	6	Petermann	
22	Baillaud B	84	Euctemon N	Democritus		Petermann B	25
23	Arnold H	85	Mitchell B	Democritus A	60	Petermann C	29
24	Petermann E	86	Mitchell E	Democritus B	66	Petermann D	11
25	Petermann B	87	Egede P	Democritus BA	92	Petermann E	24
26	Meton G	88	Aristoteles D	Democritus D	57	Petermann R	18
27	Meton D	89	Gärtner C	Democritus L	55	Petermann S	16
28	Arnold E	90	Schwabe N	Democritus M	54	Peters	
29	Petermann C	91	Archytas DA	Democritus N	53	Protagoras ϵ	94
30	Meton B	92	Democritus BA	Egede		Rima Aristoteles I	
31	Arnold K	93	Gärtner FA	Egede A	76	Rima Aristoteles II	
32	Arnold N	94	Protagoras ϵ	Egede B	77	Rima Sheepshanks	
33	Meton C	95	Galle BA	Egede C	78	Schwabe C	38
34	Arnold L	96	Aristoteles γ	Egede E	81	Schwabe G	48
35	Schwabe W	97	Aristoteles θ	Egede F	75	Schwabe N	90
36	Arnold A	98	Aristoteles ζ	Egede M	79	Schwabe W	35
37	Arnold M	99	Mitchell A	Egede N	80	Sheepshanks	
38	Schwabe C	100	Arnold κ	Egede P	87	Sheepshanks A	64
39	Arnold F	101	Kane A	Euctemon		Sheepshanks B	63
40	Neison D	102	Arnold J	Euctemon C	13	Sheepshanks C	70

Cusanus 1
2
3 4
5
6
7
10
8
9
11 12
13
14 15
16
17
Schwabe
18
19 20
Hayn
25
22
24
26
21
23
27
29
30
28
31
32
33
34 35
Strabo
36
Thales
37
38
39
40
Gärtner
41
42
De La Rue
I Rima Gärtner
44 45
68
43
46
66
47
67
48
49
50
51
53
52
Mare Frigoris
54
55
56
57
58
60
Keldysh
59
61
62
Baily
63
64
65
Hercules

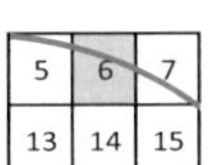

6

No.	Name	No.	Name	Name	No.	Name	No.
	Cusanus	50	De La Rue E	Aristoteles M	54	Schwabe E	21
	Schwabe	51	Endymion G	Aristoteles N	56	Schwabe F	16
	Hayn	52	Gärtner M	Atlas E	64	Schwabe G	18
	Strabo	53	De La Rue W	Atlas G	61	Schwabe K	12
	Thales	54	Aristoteles M	Baily		Schwabe U	17
	Gärtner	55	Endymion J	Baily A	63	Schwabe W	6
	Rima Gärtner	56	Aristoteles N	Baily B	59	Schwabe X	9
	Rima Gärtner I	57	Endymion X	Baily K	58	Strabo	
	De La Rue	58	Baily K	Cusanus		Strabo B	24
	Mare Frigoris	59	Baily B	Cusanus A	4	Strabo C	13
	Baily	60	Hercules H	Cusanus B	5	Strabo L	23
	Keldysh	61	Atlas G	Cusanus C	3	Strabo N	25
	Hercules	62	Hercules F	Cusanus E	1	Thales	
1	Cusanus E	63	Baily A	Cusanus F	2	Thales A	45
2	Cusanus F	64	Atlas E	Cusanus H	7	Thales E	48
3	Cusanus C	65	Hercules B	De La Rue		Thales F	42
4	Cusanus A	66	Gärtner FA	De La Rue D	49	Thales G	35
5	Cusanus B	67	De La Rue EA	De La Rue E	50	Thales H	38
6	Schwabe W	68	De La Rue J	De La Rue EA	67	Thales W	44
7	Cusanus H			De La Rue J	68		
8	Hayn F			De La Rue P	39		
9	Schwabe X			De La Rue Q	36		
10	Hayn S			De La Rue R	33		
11	Schwabe C			De La Rue S	31		
12	Schwabe K			De La Rue W	53		
13	Strabo C			Democritus K	30		
14	Hayn J			Democritus L	29		
15	Hayn E			Endymion B	40		
16	Schwabe F			Endymion C	46		
17	Schwabe U			Endymion G	51		
18	Schwabe G			Endymion J	55		
19	Hayn D			Endymion X	57		
20	Hayn B			Gärtner			
21	Schwabe E			Gärtner A	37		
22	Schwabe D			Gärtner D	43		
23	Strabo L			Gärtner E	34		
24	Strabo B			Gärtner F	47		
25	Strabo N			Gärtner FA	66		
26	Hayn L			Gärtner G	41		
27	Hayn H			Gärtner M	52		
28	Hayn A			Hayn			
29	Democritus L			Hayn A	28		
30	Democritus K			Hayn B	20		
31	De La Rue S			Hayn D	19		
32	Hayn M			Hayn E	15		
33	De La Rue R			Hayn F	8		
34	Gärtner E			Hayn H	27		
35	Thales G			Hayn J	14		
36	De La Rue Q			Hayn L	26		
37	Gärtner A			Hayn M	32		
38	Thales H			Hayn S	10		
39	De La Rue P			Hercules			
40	Endymion B			Hercules B	65		
41	Gärtner G			Hercules F	62		
42	Thales F			Hercules H	60		
43	Gärtner D			Keldysh			
44	Thales W			Mare Frigoris			
45	Thales A			Rima Gärtner			
46	Endymion C			Rima Gärtner I			
47	Gärtner F			Schwabe			
48	Thales E			Schwabe C	11		
49	De La Rue D			Schwabe D	22		

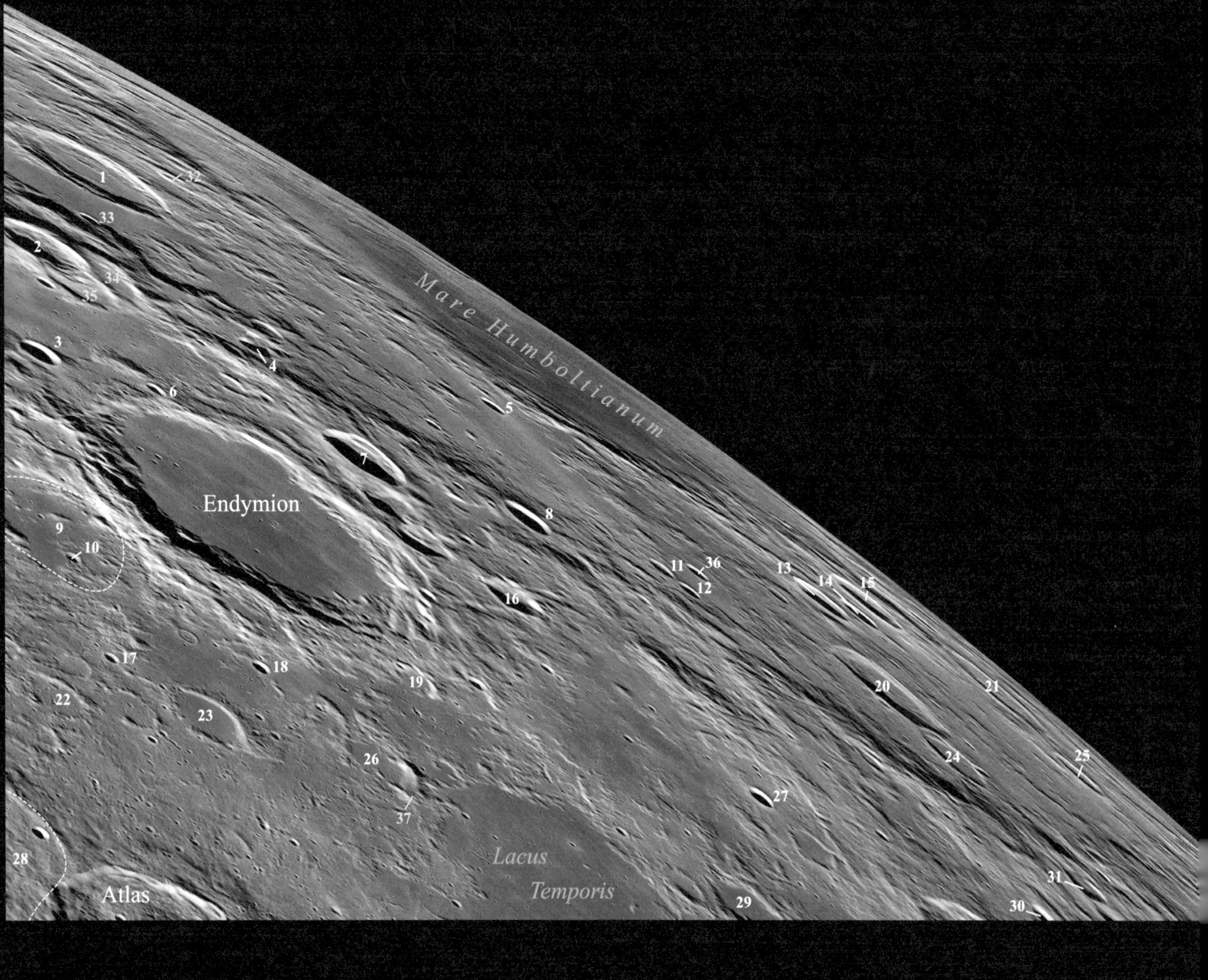

Mare Humboltianum

Endymion

Atlas

Lacus Temporis

1 2 3 4 5 6 7 8 9 10 11 12 13 14 15 16 17 18 19 20 21 22 23 24 25 26 27 28 29 30 31 32 33 34 35 36 37

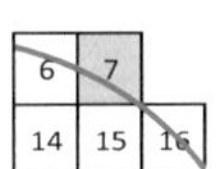

7

No.	Name	No.	Name	Name	No.	Name	No.
	Mare Humboldtianum			Atlas			
	Endymion			Atlas D	23		
	Atlas			Atlas E	28		
	Lacus Temporis			Atlas G	22		
1	Endymion B			Atlas L	17		
2	Endymion C			Atlas P	26		
3	Endymion G			Atlas ϵ	37		
4	Endymion F			Boss A	15		
5	Endymion L			Boss B	14		
6	Endymion Y			Boss C	13		
7	Endymion A			Boss K	25		
8	Endymion E			Boss L	21		
9	Endymion J			Endymion			
10	Endymion X			Endymion A	7		
11	Endymion W			Endymion B	1		
12	Endymion N			Endymion BA	32		
13	Boss C			Endymion BB	33		
14	Boss B			Endymion C	2		
15	Boss A			Endymion CA	35		
16	Endymion D			Endymion CB	34		
17	Atlas L			Endymion D	16		
18	Endymion K			Endymion E	8		
19	Endymion H			Endymion F	4		
20	Mercurius M			Endymion G	3		
21	Boss L			Endymion H	19		
22	Atlas G			Endymion J	9		
23	Atlas D			Endymion K	18		
24	Mercurius E			Endymion L	5		
25	Boss K			Endymion M	36		
26	Atlas P			Endymion N	12		
27	Mercurius H			Endymion W	11		
28	Atlas E			Endymion X	10		
29	Mercurius C			Endymion Y	6		
30	Mercurius B			Lacus Temporis			
31	Mercurius A			Mare Humboldtianum			
32	Endymion BA			Mercurius A	31		
33	Endymion BB			Mercurius B	30		
34	Endymion CB			Mercurius C	29		
35	Endymion CA			Mercurius E	24		
36	Endymion M			Mercurius H	27		
37	Atlas ϵ			Mercurius M	20		

1
2
3
4
5
6
7
8
9
Dechen
10
11
12
Gerard
13
14
15
Harding
16
17
18
19
20
21
Bunsen
22
23
Von Braun
24
60
25
61
Mons
Rümker
26
62
63
27
28
29
30
Oceanus
32
Lavoisier
31
33
36
34
35
39
37
38
40
41
42
44
43
45
Naumann
48
Procellarum
Dorsum Scilla
47
52
Ulugh Beigh
51
50
53
Aston
49
64
65
54
55
Lichtenberg
56
Nielsen
57
58
Dorsa Whiston
Humason
Rima
Cleopatra
Rima
Agricola
Rimae
Aristarchus
Wollaston
Aloha
59
66

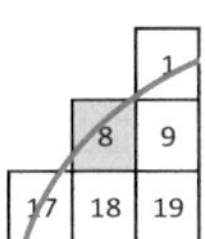

No.	Name	No.	Name	Name	No.	Name	No.
	Dechen	41	Lavoisier Z	Aloha		Repsold S	3
	Gerard	42	Ulugh Beigh M	Aston		Repsold T	2
	Harding	43	Lavoisier C	Aston K	45	Rima Agricola	
	Bunsen	44	Aston L	Aston L	44	Rima Cleopatra	
	Von Braun	45	Aston K	Briggs η	66	Rimae Aristarchus	
	Mons Rümker	47	Ulugh Beigh A	Bunsen		Rümker C	25
	Lavoisier	48	Lichtenberg R	Bunsen D	22	Rümker E	32
	Naumann	49	Ulugh Beigh B	Dechen		Rümker F	35
	Oceanus Procellarum	50	Lichtenberg F	Dechen A	9	Rümker S	19
	Ulugh Beigh	51	Lichtenberg B	Dechen B	14	Rümker T	18
	Aston	52	Naumann G	Dechen C	8	Rümker α	60
	Lichtenberg	53	Wollaston D	Dorsa Whiston		Rümker β	63
	Nielsen	54	Ulugh Beigh D	Dorsum Scilla		Rümker ζ	61
	Humason	55	Ulugh Beigh C	Gerard		Rümker θ	62
	Aloha	56	Lichtenberg H	Gerard A	10	Ulugh Beigh	
	Wollaston	57	Wollaston V	Gerard C	7	Ulugh Beigh A	47
	Rima Agricola	58	Wollaston U	Gerard D	6	Ulugh Beigh B	49
	Rimae Aristarchus	59	Herodotus E	Gerard E	11	Ulugh Beigh C	55
	Montes Agricola	60	Rümker α	Gerard F	12	Ulugh Beigh D	54
	Dorsum Scilla	61	Rümker ζ	Gerard K	13	Ulugh Beigh M	42
	Rima Cleopatra	62	Rümker θ	Gerard L	15	Von Braun	
	Dorsa Whiston	63	Rümker β	Gerard Q Inner	5	Wollaston	
1	Repsold C	64	Lichtenberg β	Gerard Q Outer	4	Wollaston D	53
2	Repsold T	65	Lichtenberg ϵ	Harding		Wollaston U	58
3	Repsold S	66	Briggs η	Harding A	26	Wollaston V	57
4	Gerard Q Outer			Harding B	21		
5	Gerard Q Inner			Harding C	17		
6	Gerard D			Harding D	16		
7	Gerard C			Harding H	24		
8	Dechen C			Herodotus E	59		
9	Dechen A			Humason			
10	Gerard A			Lavoisier			
11	Gerard E			Lavoisier A	38		
12	Gerard F			Lavoisier B	27		
13	Gerard K			Lavoisier C	43		
14	Dechen B			Lavoisier E	23		
15	Gerard L			Lavoisier F	37		
16	Harding D			Lavoisier G	36		
17	Harding C			Lavoisier H	31		
18	Rümker T			Lavoisier J	33		
19	Rümker S			Lavoisier K	29		
20	Lavoisier N			Lavoisier L	28		
21	Harding B			Lavoisier N	20		
22	Bunsen D			Lavoisier S	30		
23	Lavoisier E			Lavoisier T	40		
24	Harding H			Lavoisier W	39		
25	Rümker C			Lavoisier Z	41		
26	Harding A			Lichtenberg			
27	Lavoisier B			Lichtenberg B	51		
28	Lavoisier L			Lichtenberg F	50		
29	Lavoisier K			Lichtenberg H	56		
30	Lavoisier S			Lichtenberg R	48		
31	Lavoisier H			Lichtenberg β	64		
32	Rümker E			Lichtenberg ϵ	65		
33	Lavoisier J			Mons Rümker			
34	Naumann B			Montes Agricola			
35	Rümker F			Naumann			
36	Lavoisier G			Naumann B	34		
37	Lavoisier F			Naumann G	52		
38	Lavoisier A			Nielsen			
39	Lavoisier W			Oceanus Procellarum			
40	Lavoisier T			Repsold C	1		

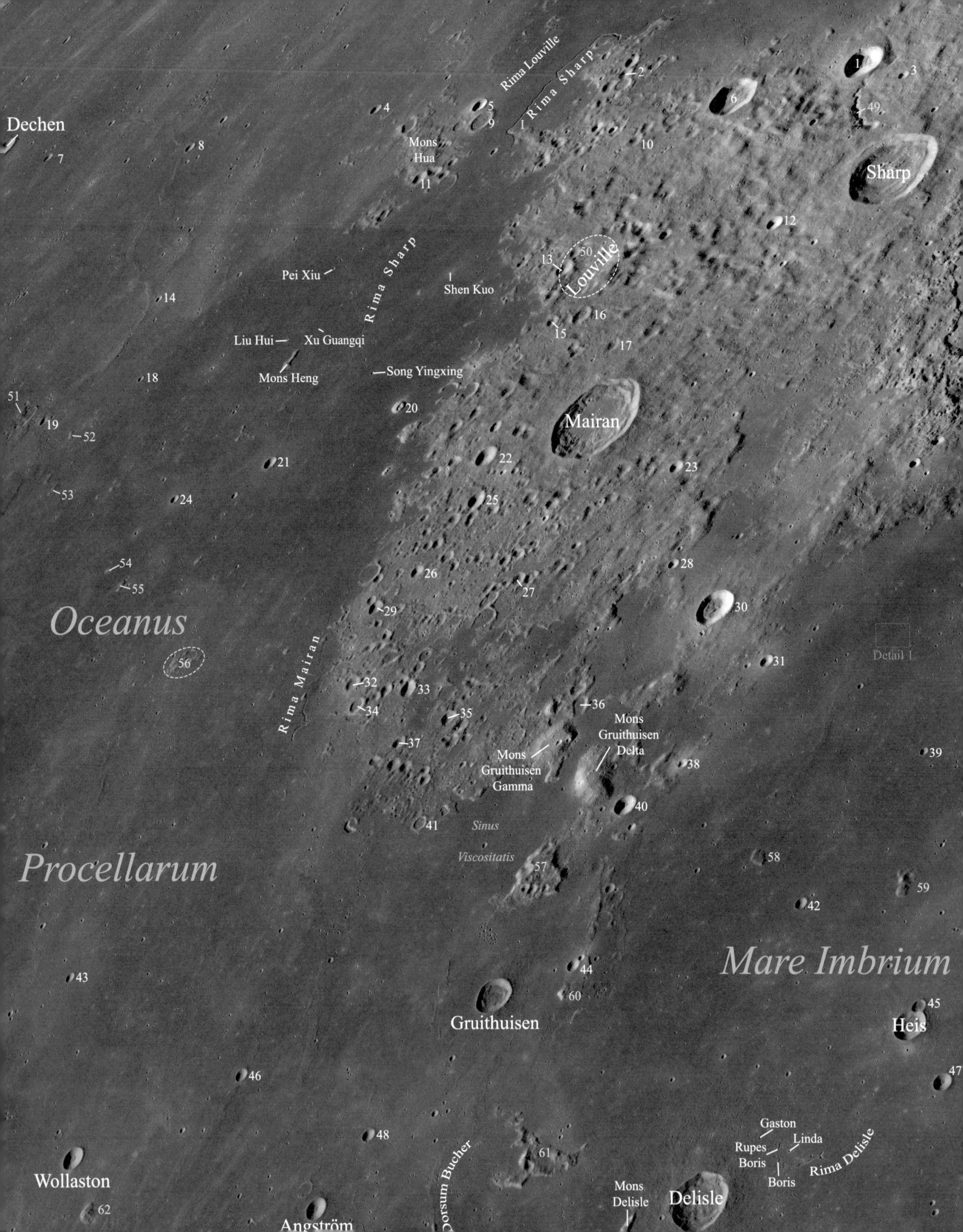

Dechen
7
8
4
5
9
Mons
Hua
11
Rima Louville
Rima Sharp
1
2
3
6
49
10
Sharp
12
13
50
Louville
14
Pei Xiu
Rima Sharp
Shen Kuo
16
15
17
Liu Hui
Xu Guangqi
Mons Heng
Song Yingxing
18
51
19
52
20
Mairan
21
22
23
53
24
25
54
55
26
27
28
29
30
Oceanus
56
Rima Mairan
Detail 1
31
32
33
34
35
36
Mons
Gruithuisen
Delta
37
Mons
Gruithuisen
Gamma
38
39
40
41
Sinus
Viscositatis
57
58
Procellarum
59
42
43
Mare Imbrium
44
60
45
Gruithuisen
Heis
46
47
48
Gaston
Linda
Rupes
Boris
Boris
Rima Delisle
61
Dorsum Bucher
Wollaston
Mons
Delisle
Delisle
62
Angström

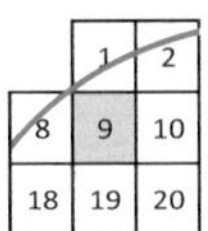

9

No.	Name
	Rima Louville
	Rima Sharp
	Rima Sharp I
	Dechen
	Mons Hua
	Sharp
	Louville
	Pei Xiu
	Shen Kuo
	Liu Hui
	Xu Guangqi
	Mons Heng
	Song Yingxing
	Mairan
	Rima Mairan
	Mons Gruithuisen Gamma
	Sinus Viscositatis
	Mons Gruithuisen Delta
	Gruithuisen
	Heis
	Gaston
	Linda
	Rupes Boris
	Boris
	Rima Delisle
	Wollaston
	Delisle
	Angström
	Mons Delisle
	Dorsum Bucher
1	Sharp A
2	Sharp U
3	Sharp M
4	Louville K
5	Louville D
6	Sharp B
7	Dechen A
8	Dechen D
9	Louville DA
10	Sharp V
11	Louville P
12	Sharp D
13	Louville B
14	Rümker L
15	Louville E
16	Louville A
17	Mairan Y
18	Rümker K
19	Rümker C
20	Mairan T
21	Mairan G
22	Mairan D
23	Mairan K
24	Rümker H
25	Mairan F
26	Mairan N
27	Mairan L
28	Mairan H
29	Mairan C
30	Mairan A
31	Mairan E
32	Gruithuisen S
33	Gruithuisen E
34	Gruithuisen R
35	Gruithuisen M
36	Gruithuisen P
37	Gruithuisen G
38	Gruithuisen F
39	C. Herschel V
40	Gruithuisen B
41	Gruithuisen K
42	C. Herschel E
43	Wollaston D
44	Gruithuisen H
45	Heis A
46	Angström B
47	Heis D
48	Angström A
49	Sharp C
50	Louville B
51	Rümker α
52	Rümker ζ
53	Rümker θ
54	Rümker η
55	Rümker ξ
56	Wollaston γ
57	Gruithuisen ζ
58	C. Herschel ϵ
59	C. Herschel ζ
60	Delisle γ
61	Delisle α
62	Wollaston α

Name	No.
Angström	
Angström A	48
Angström B	46
Boris	
C. Herschel E	42
C. Herschel V	39
C. Herschel ϵ	58
C. Herschel ζ	59
Dechen	
Dechen A	7
Dechen D	8
Delisle	
Delisle α	61
Delisle γ	60
Dorsum Bucher	
Gaston	
Gruithuisen	
Gruithuisen B	40
Gruithuisen E	33
Gruithuisen F	38
Gruithuisen G	37
Gruithuisen H	44
Gruithuisen K	41
Gruithuisen M	35
Gruithuisen P	36
Gruithuisen R	34
Gruithuisen S	32
Gruithuisen ζ	57
Heis	
Heis A	45
Heis D	47
Linda	
Liu Hui	
Louville	
Louville A	16
Louville B	13
Louville B	50
Louville D	5
Louville DA	9
Louville E	15
Louville K	4
Louville P	11
Mairan	
Mairan A	30
Mairan C	29
Mairan D	22
Mairan E	31
Mairan F	25
Mairan G	21
Mairan H	28
Mairan K	23
Mairan L	27
Mairan N	26
Mairan T	20
Mairan Y	17
Mons Delisle	
Mons Gruithuisen Delta	
Mons Gruithuisen Gamma	
Mons Heng	
Mons Hua	
Pei Xiu	
Rima Delisle	
Rima Louville	
Rima Mairan	
Rima Sharp	
Rima Sharp I	
Rümker C	19
Rümker H	24
Rümker K	18
Rümker L	14
Rümker α	51
Rümker ζ	52
Rümker η	54
Rümker θ	53
Rümker ξ	55
Rupes Boris	
Sharp	
Sharp A	1
Sharp B	6
Sharp C	49
Sharp D	12
Sharp M	3
Sharp U	2
Sharp V	10
Shen Kuo	
Sinus Viscositatis	
Song Yingxing	
Wollaston	
Wollaston D	43
Wollaston α	62
Wollaston γ	56
Xu Guangqi	

1
8
9
Bianchini
7
2
3
4
6
36
5
10
12
Sharp
11
Promontorium
Laplace
Sinus Iridum
13
14
15
Promontorium
Heraclides
16
17a
18
Le Verrier
17
Helicon
19
20
21
22
23
24
Mare Imbrium
37
25
26
27
28
C. Herschel
38
Carlini
29
Heis
30
31
32
Dorsum Heim
35
33
34
Dorsum Zirkel
McDonald
Caventou

1	2	3
9	10	11
19	20	21

10

No.	Name	No.	Name
	[illegible]		
	Bianchini		
	Montes Jura		
	Sinus Iridum		
	Promontorium Laplace		
	Promontorium Heraclides		
	Helicon		
	Le Verrier		
	C. Herschel		
	Carlini		
	Heis		
	Dorsum Heim		
	Dorsum Zirkel		
	Caventou		
	McDonald		
	Mare Imbrium		
1	Bianchini W		
2	Sharp A		
3	Sharp M		
4	Sharp K		
5	Sharp J		
6	Bianchini D		
7	Bianchini H		
8	Bianchini N		
9	Bianchini M		
10	Laplace D		
11	Sharp L		
12	Bianchini G		
13	Laplace A		
14	Heraclides E		
15	Helicon G		
16	Heraclides A		
17	Helicon C		
18	Helicon E		
19	Le Verrier T		
20	Le Verrier S		
21	Heraclides F		
22	Carlini S		
23	Helicon B		
24	C. Herschel C		
25	C. Herschel V		
26	C. Herschel U		
27	Carlini A		
28	Carlini C		
29	Heis A		
30	Carlini G		
31	Carlini H		
32	Heis D		
33	Carlini L		
34	Carlini K		
35	Carlini E		
36	Sharp C		
37	Helicon BA		
38	C. Herschel ζ		
17a	Helicon C*		

Name	No.	Name	No
[illegible]			
Bianchini D	6		
Bianchini G	12		
Bianchini H	7		
Bianchini M	9		
Bianchini N	8		
Bianchini W	1		
C. Herschel			
C. Herschel C	24		
C. Herschel U	26		
C. Herschel V	25		
C. Herschel ζ	38		
Carlini			
Carlini A	27		
Carlini C	28		
Carlini E	35		
Carlini G	30		
Carlini H	31		
Carlini K	34		
Carlini L	33		
Carlini S	22		
Caventou			
Dorsum Heim			
Dorsum Zirkel			
Heis			
Heis A	29		
Heis D	32		
Helicon			
Helicon B	23		
Helicon BA	37		
Helicon C	17		
Helicon C*	17a		
Helicon E	18		
Helicon G	15		
Heraclides A	16		
Heraclides E	14		
Heraclides F	21		
Laplace A	13		
Laplace D	10		
Le Verrier			
Le Verrier S	20		
Le Verrier T	19		
Mare Imbrium			
McDonald			
Montes Jura			
Promontorium Heraclides			
Promontorium Laplace			
Sharp			
Sharp A	2		
Sharp C	36		
Sharp J	5		
Sharp K	4		
Sharp L	11		
Sharp M	3		
Sinus Iridum			

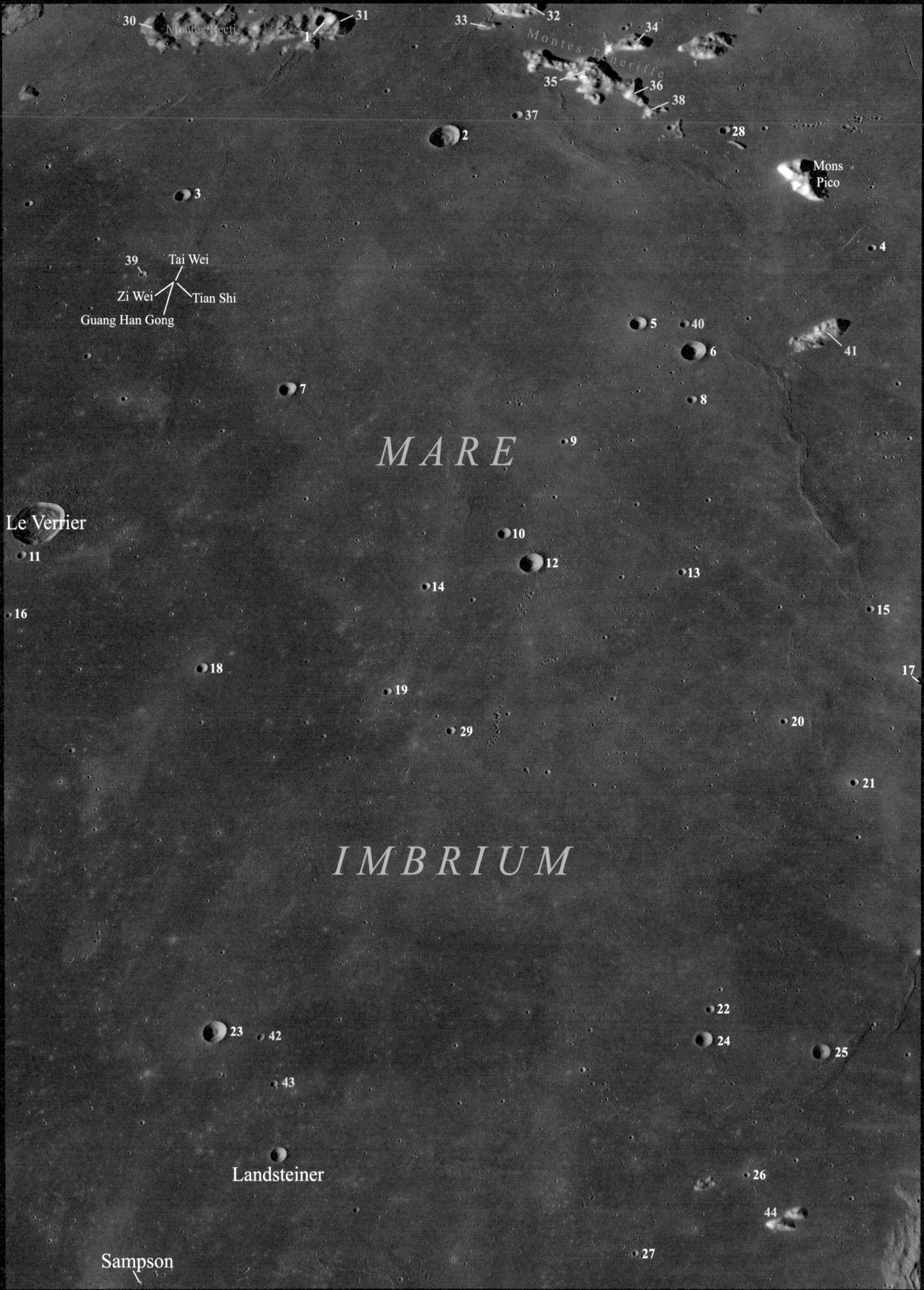

30 Montes Recti 1 31
33 32
Montes Teneriffe 34
35 36 38
37
2 28
Mons Pico
3
4
39 Tai Wei
Zi Wei Tian Shi
Guang Han Gong
5 40
6 41
7
8
MARE
9
Le Verrier
10
11
12
13
14
15
16
18
17
19
20
29
21
IMBRIUM
22
23 42
24
25
43
Landsteiner
26
44
27
Sampson

No.	Name	No.	Name	Name	No.	Name	No.
	Montes Recti			Archimedes X	26		
	Montes Teneriffe			Archimedes Y	27		
	Mons Pico			Archimedes ζ	44		
	Tai Wei			Carlini D	23		
	Zi Wei			Carlini DA	43		
	Tian Shi			Carlini DB	42		
	Guang Han Gong			Guang Han Gong			
	Mare Imbrium			Kirch E	21		
	Le Verrier			Kirch F	17		
	Landsteiner			Kirch G	20		
	Sampson			Kirch H	15		
1	Montes Recti B			Kirch M	13		
2	Pico B			Landsteiner			
3	Laplace F			Laplace F	3		
4	Pico K			Laplace FA	39		
5	Pico D			Le Verrier			
6	Pico E			Le Verrier A	18		
7	Le Verrier E			Le Verrier B	10		
8	Pico F			Le Verrier D	12		
9	Le Verrier X			Le Verrier E	7		
10	Le Verrier B			Le Verrier S	16		
11	Le Verrier T			Le Verrier T	11		
12	Le Verrier D			Le Verrier U	29		
13	Kirch M			Le Verrier V	19		
14	Le Verrier W			Le Verrier W	14		
15	Kirch H			Le Verrier X	9		
16	Le Verrier S			Mare Imbrium			
17	Kirch F			Mons Pico			
18	Le Verrier A			Montes Recti			
19	Le Verrier V			Montes Recti B	1		
20	Kirch G			Montes Teneriffe			
21	Kirch E			Pico B	2		
22	Spitzbergen D			Pico BA	37		
23	Carlini D			Pico D	5		
24	Spitzbergen C			Pico E	6		
25	Spitzbergen A			Pico EA	40		
26	Archimedes X			Pico F	8		
27	Archimedes Y			Pico G	28		
28	Pico G			Pico K	4		
29	Le Verrier U			Pico β	41		
30	Recti ϵ			Recti β	31		
31	Recti β			Recti ϵ	30		
32	Teneriffe ϵ			Sampson			
33	Teneriffe ω			Spitzbergen A	25		
34	Teneriffe α			Spitzbergen C	24		
35	Teneriffe δ			Spitzbergen D	22		
36	Teneriffe κ			Tai Wei			
37	Pico BA			Teneriffe α	34		
38	Teneriffe γ			Teneriffe γ	38		
39	Laplace FA			Teneriffe δ	35		
40	Pico EA			Teneriffe ϵ	32		
41	Pico β			Teneriffe κ	36		
42	Carlini DB			Teneriffe ω	33		
43	Carlini DA			Tian Shi			
44	Archimedes ζ			Zi Wei			

Vallis Alpes
Montes Alpes
4
1
62
3
2
41
40
5
Mont Blanc
6
8
7
9
19
11
Mare
10
60
38
42
Promontorium Deville
18
59
12
17
14
20
16
Promontorium Agassiz
13
Piazzi Smyth
15
21
58
23
31
22
30
Mons Piton
29
Cassini
27
28
25
24
Kirch
49
26
Imbrium
Palus Nebularum
Rima Theaetetus III
56
Theaetetus
57
44
61
45
Montes Spitzbergen
32
46
Aristillus
47
48
36
Rima Theaetetus II
33
34
Rimae Theaetetus
55
Rima Theaetetus I
35
Sinus Lunicus
43
37
51
50
39
Autolycus
52
53
Archimedes
54

12

No.	Name
	Vallis Alpes
	Montes Alpes
	Mont Blanc
	Mare Imbrium
	Promontorium Deville
	Piazzi Smyth
	Promontorium Agassiz
	Mons Piton
	Cassini
	Kirch
	Palus Nebularum
	Rima Theaetetus III
	Theaetetus
	Montes Spitzbergen
	Aristillus
	Rimae Theaetetus
	Rima Theaetetus I
	Rima Theaetetus II
	Sinus Lunicus
	Autolycus
	Archimedes
1	Pico C
2	Plato KA
3	Plato K
4	Trouvelot G
5	Alpes B
6	Piazzi Smyth M
7	Cassini P
8	Cassini K
9	Cassini G
10	Cassini L
11	Cassini X
12	Cassini E
13	Cassini C
14	Cassini W
15	Cassini M
16	Cassini Y
17	Piazzi Smyth W
18	Piazzi Smyth Y
19	Pico K
20	Piazzi Smyth Z
21	Piazzi Smyth V
22	Piazzi Smyth B
23	Piazzi Smyth U
24	Kirch H
25	Kirch K
26	Kirch F
27	Piton A
28	Piton B
29	Cassini B
30	Cassini A
31	Cassini F
32	Aristillus B
33	Archimedes V
34	Archimedes U
35	Archimedes D
36	Aristillus A
37	Archimedes C
38	Cassini Z
39	Autolycus A
40	Trouvelot η
41	Trouvelot ξ
42	Cassini N
43	Archimedes ϵ
44	Spitzbergen κ
45	Spitzbergen β
46	Spitzbergen δ
47	Spitzbergen γ
48	Spitzbergen α
49	Piton γ
50	Autolycus K
51	Autolycus η
52	Autolycus ω
53	Archimedes T
54	Archimedes S
55	Archimedes ξ
56	Spitzbergen μ
57	Kirch E
58	Piazzi Smyth β
59	Piazzi Smyth α
60	Piazzi Smyth π
61	Spitzbergen ϵ
62	Plato KB

Name	No.
Alpes B	5
Archimedes	
Archimedes C	37
Archimedes D	35
Archimedes S	54
Archimedes T	53
Archimedes U	34
Archimedes V	33
Archimedes ϵ	43
Archimedes ξ	55
Aristillus	
Aristillus A	36
Aristillus B	32
Autolycus	
Autolycus A	39
Autolycus K	50
Autolycus η	51
Autolycus ω	52
Cassini	
Cassini A	30
Cassini B	29
Cassini C	13
Cassini E	12
Cassini F	31
Cassini G	9
Cassini K	8
Cassini L	10
Cassini M	15
Cassini N	42
Cassini P	7
Cassini W	14
Cassini X	11
Cassini Y	16
Cassini Z	38
Kirch	
Kirch E	57
Kirch F	26
Kirch H	24
Kirch K	25
Mare Imbrium	
Mons Piton	
Mont Blanc	
Montes Alpes	
Montes Spitzbergen	
Palus Nebularum	
Piazzi Smyth	
Piazzi Smyth B	22
Piazzi Smyth M	6
Piazzi Smyth U	23
Piazzi Smyth V	21
Piazzi Smyth W	17
Piazzi Smyth Y	18
Piazzi Smyth Z	20
Piazzi Smyth α	59
Piazzi Smyth β	58
Piazzi Smyth π	60
Pico C	1
Pico K	19
Piton A	27
Piton B	28
Piton γ	49
Plato K	3
Plato KA	2
Plato KB	62
Promontorium Agassiz	
Promontorium Deville	
Rima Theaetetus I	
Rima Theaetetus II	
Rima Theaetetus III	
Rimae Theaetetus	
Sinus Lunicus	
Spitzbergen α	48
Spitzbergen β	45
Spitzbergen γ	47
Spitzbergen δ	46
Spitzbergen ϵ	61
Spitzbergen κ	44
Spitzbergen μ	56
Theaetetus	
Trouvelot G	4
Trouvelot η	40
Trouvelot ξ	41
Vallis Alpes	

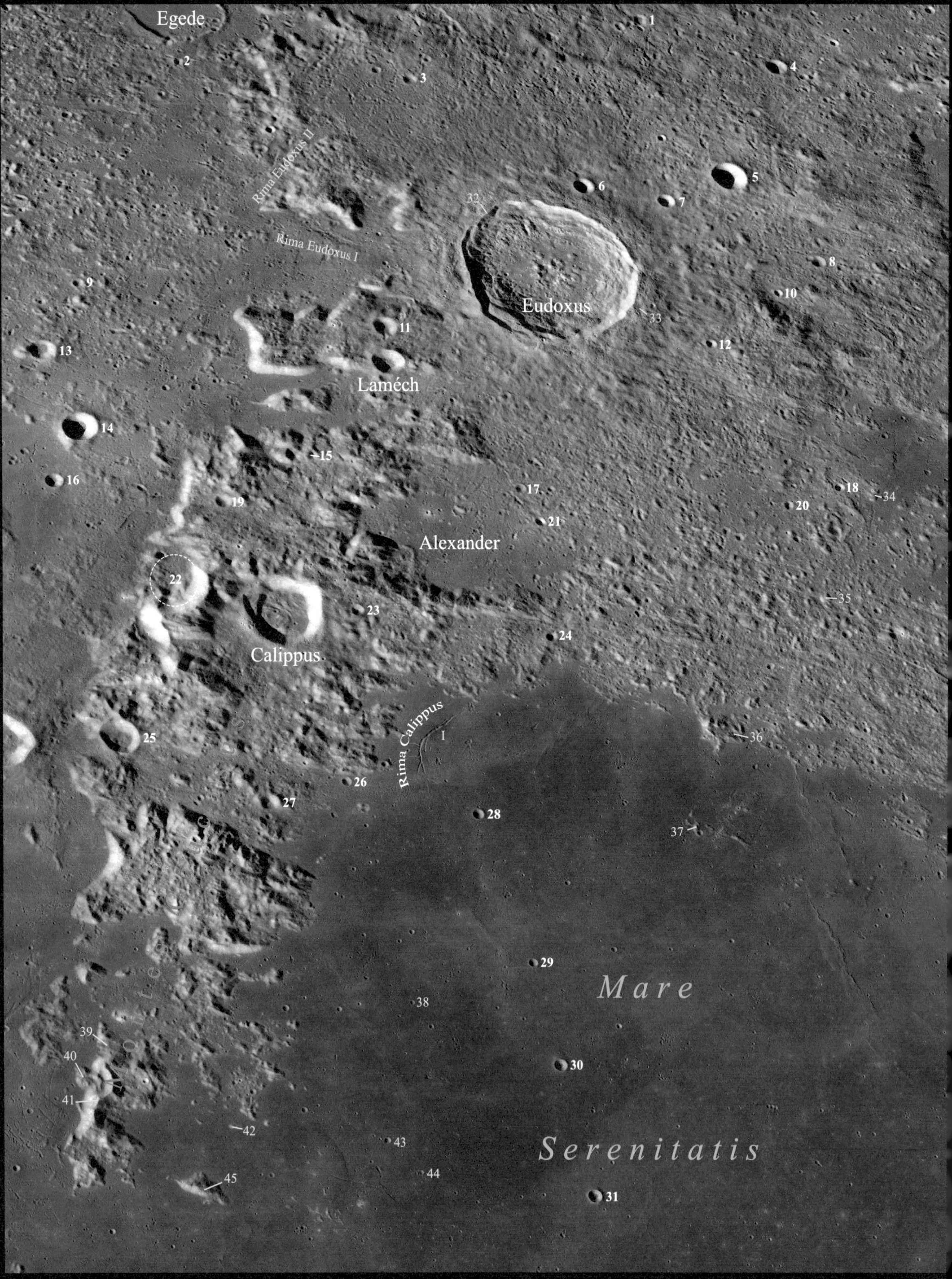

Egede

Rima Eudoxus II

Rima Eudoxus I

Eudoxus

Laméch

Alexander

Calippus

Rima Calippus I

Mare Serenitatis

1 2 3 4 5 6 7 8 9 10 11 12 13 14 15 16 17 18 19 20 21 22 23 24 25 26 27 28 29 30 31 32 33 34 35 36 37 38 39 40 41 42 43 44 45

4	5	6
12	**13**	14
22	23	24

13

No.	Name	No.	Name
	Egede		
	Rima Eudoxus II		
	Rima Eudoxus I		
	Eudoxus		
	Lamèch		
	Alexander		
	Calippus		
	Rima Calippus		
	Rima Calippus I		
	Montes Caucasus		
	Mare Serenitatis		
1	Mitchell B		
2	Egede P		
3	Aristoteles D		
4	Mitchell E		
5	Eudoxus A		
6	Eudoxus B		
7	Eudoxus G		
8	Eudoxus E		
9	Cassini X		
10	Eudoxus U		
11	Eudoxus D		
12	Eudoxus V		
13	Cassini E		
14	Cassini C		
15	Calippus G		
16	Cassini F		
17	Alexander A		
18	Eudoxus J		
19	Calippus F		
20	Alexander K		
21	Alexander B		
22	Calippus C		
23	Calippus E		
24	Alexander C		
25	Calippus A		
26	Calippus D		
27	Calippus B		
28	Linné G		
29	Linné H		
30	Linné F		
31	Linné B		
32	Eudoxus γ		
33	Eudoxus W		
34	Alexander β		
35	Alexander ψ		
36	Alexander κ		
37	Alexander ϕ		
38	Linné BC		
39	Calippus ν		
40	Calippus γ		
41	Calippus β		
42	Calippus λ		
43	Linné BB		
44	Linné BA		
45	Calippus α		

Name	No.	Name	No.
Alexander			
Alexander A	17		
Alexander B	21		
Alexander C	24		
Alexander K	20		
Alexander β	34		
Alexander κ	36		
Alexander ϕ	37		
Alexander ψ	35		
Aristoteles D	3		
Calippus			
Calippus A	25		
Calippus B	27		
Calippus C	22		
Calippus D	26		
Calippus E	23		
Calippus F	19		
Calippus G	15		
Calippus α	45		
Calippus β	41		
Calippus γ	40		
Calippus λ	42		
Calippus ν	39		
Cassini C	14		
Cassini E	13		
Cassini F	16		
Cassini X	9		
Egede			
Egede P	2		
Eudoxus			
Eudoxus A	5		
Eudoxus B	6		
Eudoxus D	11		
Eudoxus E	8		
Eudoxus G	7		
Eudoxus J	18		
Eudoxus U	10		
Eudoxus V	12		
Eudoxus W	33		
Eudoxus γ	32		
Lamèch			
Linné B	31		
Linné BA	44		
Linné BB	43		
Linné BC	38		
Linné F	30		
Linné G	28		
Linné H	29		
Mare Serenitatis			
Mitchell B	1		
Mitchell E	4		
Montes Caucasus			
Rima Calippus			
Rima Calippus I			
Rima Eudoxus I			
Rima Eudoxus II			

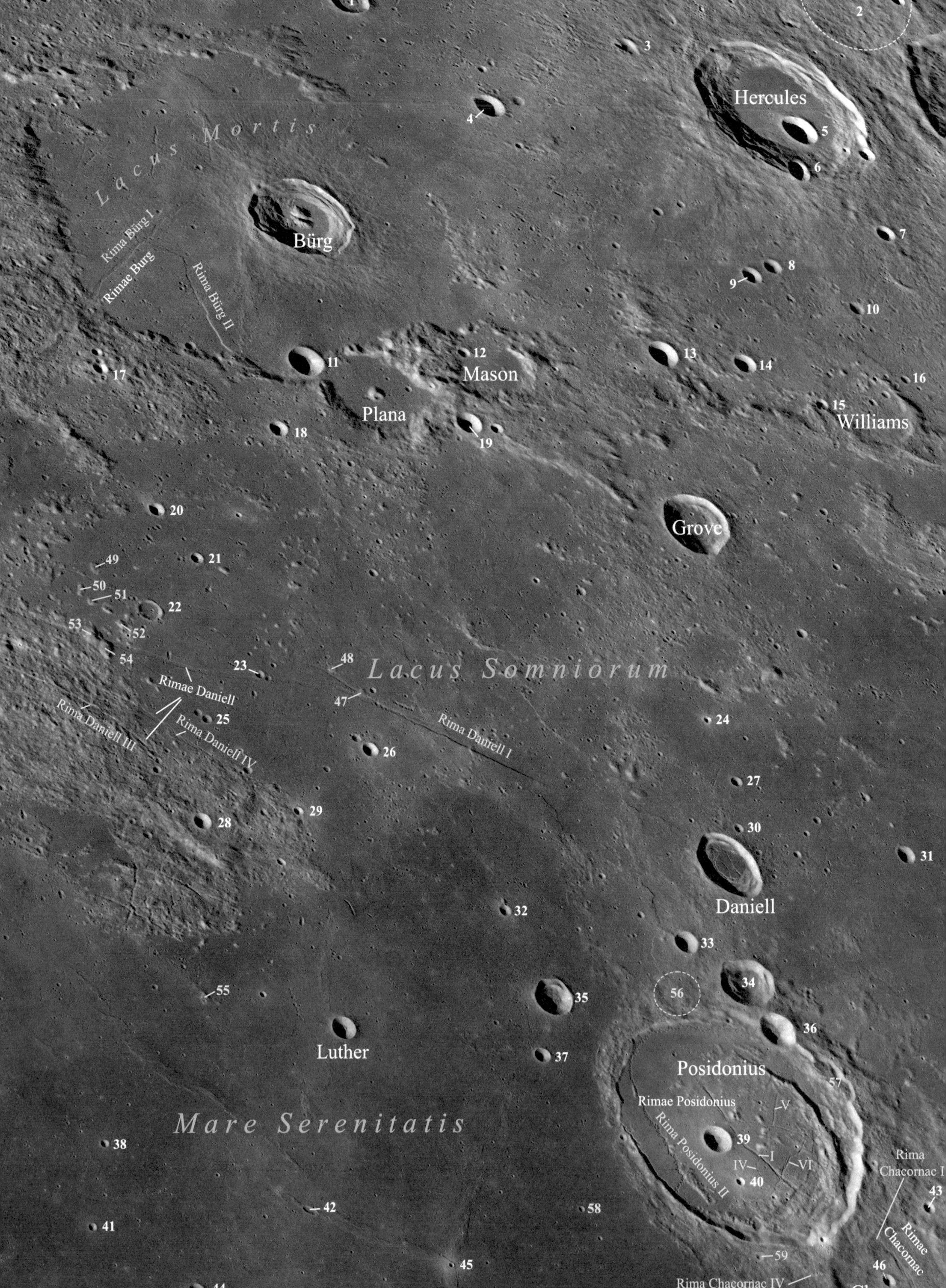
1
2
3
Hercules
5
6
4
Lacus Mortis
Bürg
Rima Bürg I
Rimae Burg
Rima Bürg II
7
8
9
10
11
12
Mason
13
14
16
17
Plana
15
Williams
18
19
20
Grove
49
21
50
51
22
53
52
54
23
48
Lacus Somniorum
Rimae Daniell
47
24
Rima Daniell III
25
Rima Daniell IV
Rima Daniell I
26
27
29
28
30
31
Daniell
32
33
34
55
35
56
36
Luther
37
Posidonius
57
Rimae Posidonius
V
Mare Serenitatis
Rima Posidonius II
39
I
IV
VI
38
40
Rima Chacornac I
43
42
58
41
Rimae Chacornac
59
45
46
Rima Chacornac IV
44
Chacornac

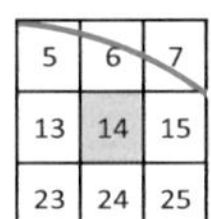

14

No.	Name
	Hercules
	Lacus Mortis
	Rimae Bürg
	Bürg
	Mason
	Plana
	Williams
	Grove
	Lacus Somniorum
	Rimae Daniell
	Rima Daniell I
	Rima Daniell IV
	Rima Daniell III
	Daniell
	Luther
	Posidonius
	Rimae Posidonius
	Rima Posidonius I
	Rima Posidonius II
	Rima Posidonius IV
	Rima Posidonius V
	Rima Posidonius VI
	Mare Serenitatis
	Rimae Chacornac
	Rima Chacornac I
	Rima Chacornac IV
	Chacornac
1	Baily A
2	Atlas E
3	Hercules B
4	Bürg A
5	Hercules G
6	Hercules E
7	Hercules D
8	Hercules K
9	Hercules J
10	Williams F
11	Plana C
12	Mason A
13	Mason C
14	Hercules C
15	Williams N
16	Williams R
17	Bürg B
18	Plana D
19	Mason B
20	Plana E
21	Plana F
22	Plana G
23	Luther Y
24	Grove Y
25	Luther K
26	Daniell D
27	Daniell X
28	Luther H
29	Luther X
30	Daniell W
31	Hall K
32	Posidonius G
33	Posidonius M
34	Posidonius J
35	Posidonius P
36	Posidonius B
37	Posidonius F
38	Posidonius W
39	Posidonius A
40	Posidonius C
41	Posidonius E
42	Posidonius Z
43	Chacornac C
44	Posidonius N
45	Posidonius Y
46	Chacornac A
47	Daniell ξ
48	Daniell η
49	Plana β
50	Plana γ
51	Plana δ
52	Plana ϵ
53	Plana ζ
54	Plana ξ
55	Luther α
56	Posidonius O
57	Posidonius D
58	Posidonius FA
59	Posidonius V

Name	No.
Atlas E	2
Baily A	1
Bürg	
Bürg A	4
Bürg B	17
Chacornac	
Chacornac A	46
Chacornac C	43
Daniell	
Daniell D	26
Daniell W	30
Daniell X	27
Daniell η	48
Daniell ξ	47
Grove	
Grove Y	24
Hall K	31
Hercules	
Hercules B	3
Hercules C	14
Hercules D	7
Hercules E	6
Hercules G	5
Hercules J	9
Hercules K	8
Lacus Mortis	
Lacus Somniorum	
Luther	
Luther H	28
Luther K	25
Luther X	29
Luther Y	23
Luther α	55
Mare Serenitatis	
Mason	
Mason A	12
Mason B	19
Mason C	13
Plana	
Plana C	11
Plana D	18
Plana E	20
Plana F	21
Plana G	22
Plana β	49
Plana γ	50
Plana δ	51
Plana ϵ	52
Plana ζ	53
Plana ξ	54
Posidonius	
Posidonius A	39
Posidonius B	36
Posidonius C	40
Posidonius D	57
Posidonius E	41
Posidonius F	37
Posidonius FA	58
Posidonius G	32
Posidonius J	34
Posidonius M	33
Posidonius N	44
Posidonius O	56
Posidonius P	35
Posidonius V	59
Posidonius W	38
Posidonius Y	45
Posidonius Z	42
Rima Chacornac I	
Rima Chacornac IV	
Rima Daniell I	
Rima Daniell III	
Rima Daniell IV	
Rima Posidonius I	
Rima Posidonius II	
Rima Posidonius IV	
Rima Posidonius V	
Rima Posidonius VI	
Rimae Bürg	
Rimae Chacornac	
Rimae Daniell	
Rimae Posidonius	
Williams	
Williams F	10
Williams N	15
Williams R	16

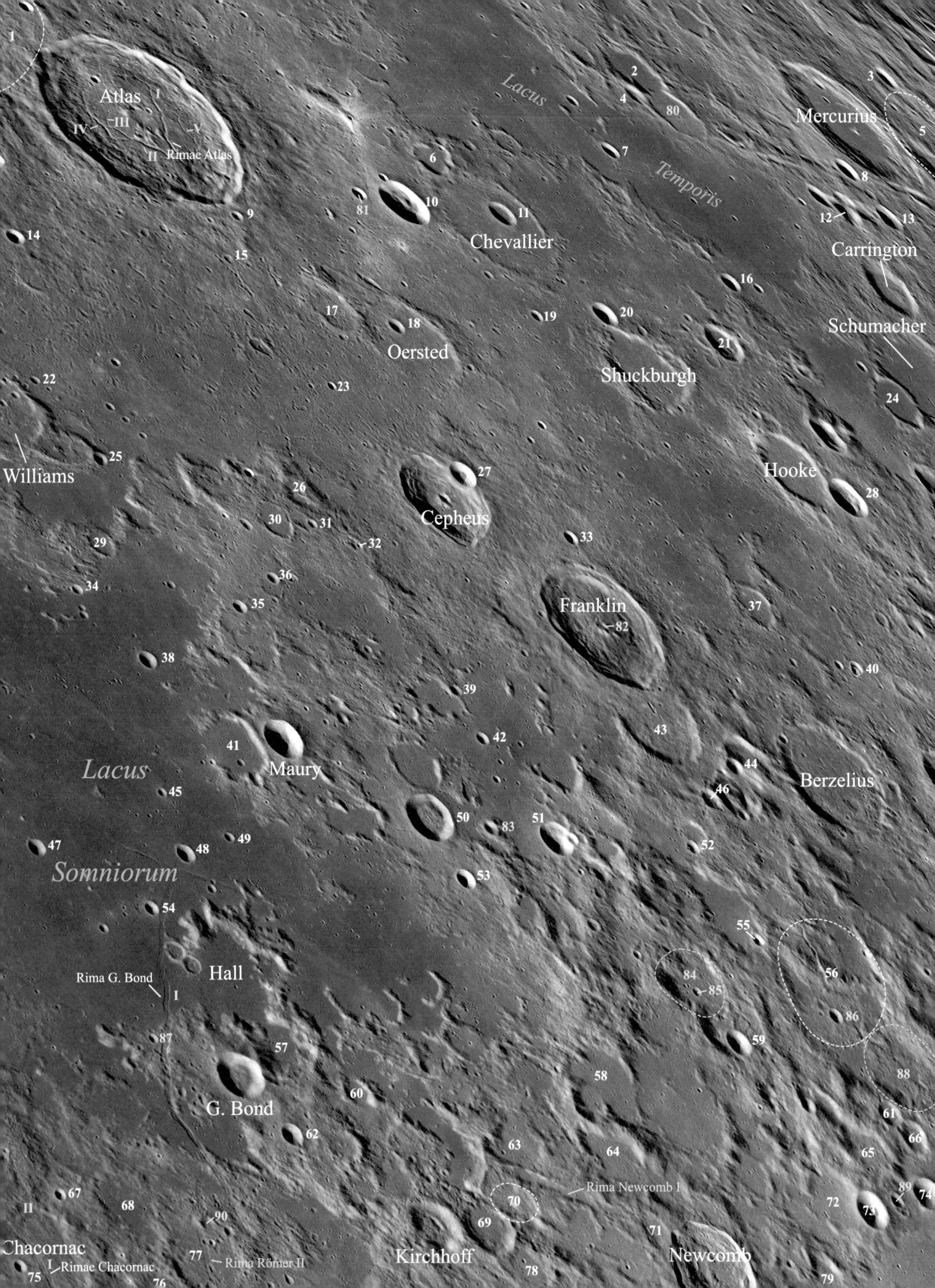

1
Atlas
I
IV
III
V
II
Rimae Atlas
Lacus
2
4
80
3
Mercurius
5
7
Temporis
6
10
81
9
11
Chevallier
8
12
13
Carrington
14
15
16
17
18
Oersted
19
20
21
Schumacher
Shuckburgh
22
23
24
Williams
25
26
27
Cepheus
Hooke
28
29
30
31
32
33
34
36
35
Franklin
82
37
38
39
40
41
Maury
42
43
44
Lacus
45
46
Berzelius
47
48
49
50
83
51
52
Somniorum
53
54
55
Rima G. Bond
I
Hall
84
85
56
86
87
57
59
58
88
G. Bond
60
61
62
63
64
65
66
67
68
70
Rima Newcomb I
72
73
89
74
II
90
69
71
Chacornac
77
Rima Römer II
Kirchhoff
Newcomb
75
I
Rimae Chacornac
76
78
79

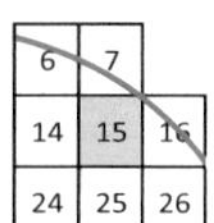

15

No.	Name	No.	Name	Name	No.	Name	No.
	Atlas	30	Maury N	Atlas		Hooke D	28
	Rimae Atlas	31	Maury L	Atlas A	10	Kirchhoff	
	Rima Atlas I	32	Maury T	Atlas AA	81	Kirchhoff C	69
	Rima Atlas II	33	Franklin G	Atlas E	1	Kirchhoff E	70
	Rima Atlas III	34	Maury U	Atlas W	15	Kirchhoff F	63
	Rima Atlas IV	35	Maury J	Atlas X	9	Kirchhoff G	78
	Rima Atlas V	36	Maury K	Berzelius		Lacus Somniorum	
	Lacus Temporis	37	Franklin K	Berzelius A	44	Lacus Temporis	
	Mercurius	38	Maury D	Berzelius B	58	Maury	
	Chevallier	39	Franklin W	Berzelius F	59	Maury A	50
	Carrington	40	Berzelius W	Berzelius FA	85	Maury AA	83
	Oersted	41	Maury C	Berzelius FB	84	Maury B	53
	Shuckburgh	42	Franklin H	Berzelius K	52	Maury C	41
	Schumacher	43	Franklin F	Berzelius T	46	Maury D	38
	Williams	44	Berzelius A	Berzelius W	40	Maury J	35
	Cepheus	45	Hall Y	Carrington		Maury K	36
	Hooke	46	Berzelius T	Cepheus		Maury L	31
	Franklin	47	Hall K	Cepheus A	27	Maury M	26
	Lacus Somniorum	48	Hall J	Chacornac		Maury N	30
	Maury	49	Hall X	Chacornac A	75	Maury P	29
	Berzelius	50	Maury A	Chacornac C	67	Maury T	32
	Hall	51	Franklin C	Chacornac D	68	Maury U	34
	Rima G. Bond	52	Berzelius K	Chacornac E	76	Mercurius	
	Rima G. Bond I	53	Maury B	Chevallier		Mercurius B	3
	G. Bond	54	Hall C	Chevallier B	11	Mercurius C	2
	Chacornac	55	Geminus W	Chevallier F	7	Mercurius CA	80
	Kirchhoff	56	Geminus E	Chevallier K	19	Mercurius D	5
	Newcomb	57	G. Bond G	Chevallier M	6	Mercurius F	12
	Rima Newcomb I	58	Berzelius B	Franklin		Mercurius G	13
	Rima Römer II	59	Berzelius F	Franklin C	51	Mercurius J	4
	Rimae Chacornac	60	G. Bond K	Franklin F	43	Mercurius L	8
	Rima Chacornac I	61	Geminus M	Franklin G	33	Newcomb	
	Rima Chacornac II	62	G. Bond A	Franklin H	42	Newcomb C	79
1	Atlas E	63	Kirchhoff F	Franklin K	37	Newcomb F	64
2	Mercurius C	64	Newcomb F	Franklin W	39	Newcomb Q	71
3	Mercurius B	65	Geminus N	Franklin γ	82	Oersted	
4	Mercurius J	66	Geminus H	G. Bond		Oersted A	18
5	Mercurius D	67	Chacornac C	G. Bond A	62	Oersted P	17
6	Chevallier M	68	Chacornac D	G. Bond B	77	Oersted U	23
7	Chevallier F	69	Kirchhoff C	G. Bond BA	90	Rima Atlas I	
8	Mercurius L	70	Kirchhoff E	G. Bond G	57	Rima Atlas II	
9	Atlas X	71	Newcomb Q	G. Bond J	87	Rima Atlas III	
10	Atlas A	72	Geminus Z	G. Bond K	60	Rima Atlas IV	
11	Chevallier B	73	Geminus D	Geminus D	73	Rima Atlas V	
12	Mercurius F	74	Geminus G	Geminus DA	89	Rima Chacornac I	
13	Mercurius G	75	Chacornac A	Geminus E	56	Rima Chacornac II	
14	Hercules D	76	Chacornac E	Geminus EA	88	Rima G. Bond	
15	Atlas W	77	G. Bond B	Geminus EB	86	Rima G. Bond I	
16	Shuckburgh E	78	Kirchhoff G	Geminus G	74	Rima Newcomb I	
17	Oersted P	79	Newcomb C	Geminus H	66	Rima Römer II	
18	Oersted A	80	Mercurius CA	Geminus M	61	Rimae Atlas	
19	Chevallier K	81	Atlas AA	Geminus N	65	Rimae Chacornac	
20	Shuckburgh C	82	Franklin γ	Geminus W	55	Schumacher	
21	Shuckburgh A	83	Maury AA	Geminus Z	72	Schumacher B	24
22	Williams R	84	Berzelius FB	Hall		Shuckburgh	
23	Oersted U	85	Berzelius FA	Hall C	54	Shuckburgh A	21
24	Schumacher B	86	Geminus EB	Hall J	48	Shuckburgh C	20
25	Williams M	87	G. Bond J	Hall K	47	Shuckburgh E	16
26	Maury M	88	Geminus EA	Hall X	49	Williams	
27	Cepheus A	89	Geminus DA	Hall Y	45	Williams M	25
28	Hooke D	90	G. Bond BA	Hercules D	14	Williams R	22
29	Maury P			Hooke			

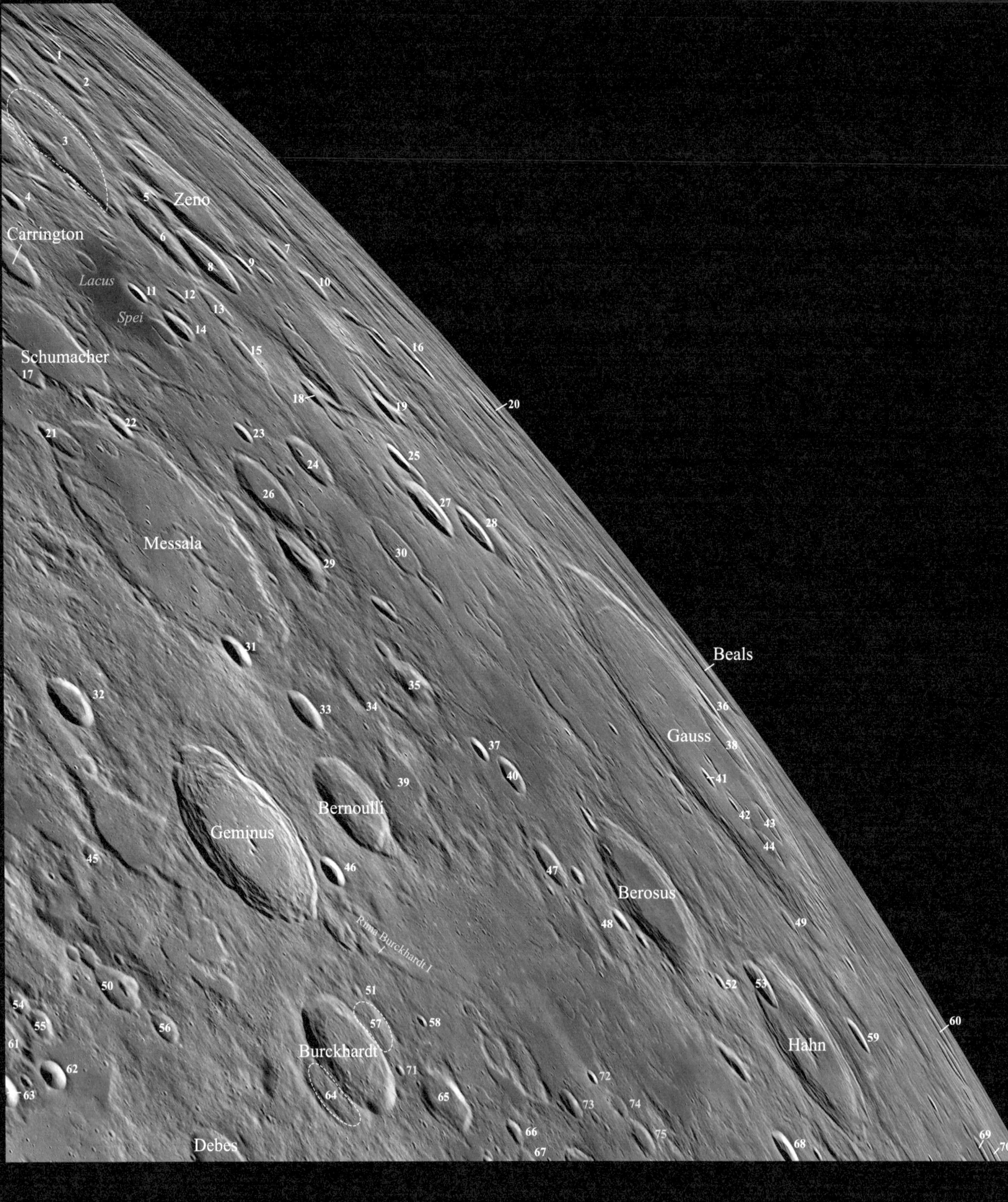
Zeno
Carrington
Lacus
Spei
Schumacher
Messala
Beals
Gauss
Geminus
Bernoulli
Berosus
Rima Burckhardt I
Burckhardt
Hahn
Debes
1
2
3
4
5
6
7
8
9
10
11
12
13
14
15
16
17
18
19
20
21
22
23
24
25
26
27
28
29
30
31
32
33
34
35
36
37
38
39
40
41
42
43
44
45
46
47
48
49
50
51
52
53
54
55
56
57
58
59
60
61
62
63
64
65
66
67
68
69
70
71
72
73
74
75

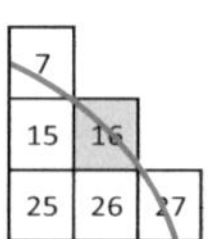

16

No.	Name	No.	Name	Name	No.	Name	No.
	[illegible]		[illegible]	[illegible]		[illegible]	
	Carrington	50	Geminus F	Bernoulli		Bernoulli	
	Lacus Spei	51	Burckhardt G	Bernoulli A	33	Bernoulli A	33
	Schumacher	52	Berosus K	Bernoulli B	35	Bernoulli B	35
	Messala	53	Hahn F	Bernoulli C	40	Bernoulli C	40
	Beals	54	Geminus M	Bernoulli D	37	Bernoulli D	37
	Gauss	55	Geminus H	Bernoulli E	39	Bernoulli E	39
	Geminus	56	Geminus A	Bernoulli K	34	Bernoulli K	34
	Bernoulli	57	Burckhardt F	Berosus		Berosus	
	Berosus	58	Burckhardt C	Berosus A	48	Berosus A	48
	Rima Burckhardt I	59	Hahn B	Berosus F	47	Berosus F	47
	Burckhardt	60	Rayleigh C	Berosus K	52	Berosus K	52
	Hahn	61	Geminus N	Burckhardt		Burckhardt	
	Debes	62	Geminus G	Burckhardt A	65	Burckhardt A	65
1	Mercurius A	63	Geminus D	Burckhardt B	66	Burckhardt B	66
2	Mercurius K	64	Burckhardt E	Burckhardt C	58	Burckhardt C	58
3	Mercurius D	65	Burckhardt A	Burckhardt E	64	Burckhardt E	64
4	Mercurius G	66	Burckhardt B	Burckhardt F	57	Burckhardt F	57
5	Zeno D	67	Cleomedes R	Burckhardt FA	71	Burckhardt FA	71
6	Zeno A	68	Hahn A	Burckhardt G	51	Burckhardt G	51
7	Zeno J	69	Seneca F	Carrington		Carrington	
8	Zeno B	70	Seneca G	Cleomedes DC	73	Cleomedes DC	73
9	Zeno G	71	Burckhardt FA	Cleomedes DE	75	Cleomedes DE	75
10	Zeno X	72	Cleomedes DG	Cleomedes DF	74	Cleomedes DF	74
11	Zeno P	73	Cleomedes DC	Cleomedes DG	72	Cleomedes DG	72
12	Zeno W	74	Cleomedes DF	Cleomedes R	67	Cleomedes R	67
13	Zeno V	75	Cleomedes DE	Debes		Debes	
14	Zeno K			Gauss		Gauss	
15	Zeno U			Gauss A	36	Gauss A	36
16	Zeno F			Gauss B	38	Gauss B	38
17	Schumacher B			Gauss C	27	Gauss C	27
18	Zeno E			Gauss D	28	Gauss D	28
19	Zeno H			Gauss E	41	Gauss E	41
20	Riemann B			Gauss F	42	Gauss F	42
21	Messala K			Gauss G	44	Gauss G	44
22	Messala J			Gauss H	49	Gauss H	49
23	Messala C			Gauss J	25	Gauss J	25
24	Messala D			Gauss W	43	Gauss W	43
25	Gauss J			Geminus		Geminus	
26	Messala E			Geminus A	56	Geminus A	56
27	Gauss C			Geminus B	45	Geminus B	45
28	Gauss D			Geminus C	46	Geminus C	46
29	Messala F			Geminus D	63	Geminus D	63
30	Messala G			Geminus F	50	Geminus F	50
31	Messala B			Geminus G	62	Geminus G	62
32	Messala A			Geminus H	55	Geminus H	55
33	Bernoulli A			Geminus M	54	Geminus M	54
34	Bernoulli K			Geminus N	61	Geminus N	61
35	Bernoulli B			Hahn		Hahn	
36	Gauss A			Hahn A	68	Hahn A	68
37	Bernoulli D			Hahn B	59	Hahn B	59
38	Gauss B			Hahn F	53	Hahn F	53
39	Bernoulli E			Lacus Spei		Lacus Spei	
40	Bernoulli C			Mercurius A	1	Mercurius A	1
41	Gauss E			Mercurius D	3	Mercurius D	3
42	Gauss F			Mercurius G	4	Mercurius G	4
43	Gauss W			Mercurius K	2	Mercurius K	2
44	Gauss G			Messala		Messala	
45	Geminus B			Messala A	32	Messala A	32
46	Geminus C			Messala B	31	Messala B	31
47	Berosus F			Messala C	23	Messala C	23
48	Berosus A			Messala D	24	Messala D	24

44
1
3
2
4
5
6
7
Russell
Briggs
8
9
10
11
45
12
13
Struve
14
15
17
16
18
19
Eddington
Seleucus
20
22
21
23
46
Oceanus
Balboa
24
25
30
26
29
Dalton
27
28
31
Krafft
32
33
41
35
34
Procellarum
37
42
36
Catena Kraft
38
Vasco da Gama
40
39
43

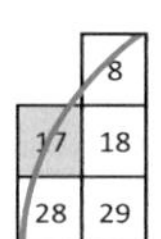

17

No.	Name
	Russell
	Briggs
	Struve
	Eddington
	Seleucus
	Balboa
	Oceanus Procellarum
	Dalton
	Krafft
	Catena Krafft
	Vasco da Gama
1	Russell S
2	Voskresenskiy K
3	Russell R
4	Russell E
5	Russell F
6	Briggs B
7	Briggs A
8	Russell B
9	Struve D
10	Struve H
11	Briggs C
12	Struve G
13	Struve K
14	Struve M
15	Struve C
16	Struve F
17	Schiaparelli A
18	Seleucus E
19	Seleucus A
20	Eddington P
21	Struve L
22	Balboa B
23	Balboa C
24	Struve B
25	Balboa D
26	Galilaei W
27	Krafft U
28	Galilaei V
29	Balboa A
30	Krafft M
31	Krafft H
32	Krafft K
33	Krafft C
34	Krafft E
35	Krafft L
36	Krafft D
37	Vasco da Gama B
38	Cardanus M
39	Cardanus K
40	Vasco da Gama F
41	Galilaei T
42	Galilaei S
43	Galilaei E
44	Briggs η
45	Briggs ξ
46	Eddington α

Name	No.
Balboa	
Balboa A	29
Balboa B	22
Balboa C	23
Balboa D	25
Briggs	
Briggs A	7
Briggs B	6
Briggs C	11
Briggs η	44
Briggs ξ	45
Cardanus K	39
Cardanus M	38
Dalton	
Eddington	
Eddington P	20
Eddington α	46
Galilaei E	43
Galilaei S	42
Galilaei T	41
Galilaei V	28
Galilaei W	26
Krafft	
Krafft C	33
Krafft D	36
Krafft E	34
Krafft H	31
Krafft K	32
Krafft L	35
Krafft M	30
Krafft U	27
Oceanus Procellarum	
Russell	
Russell B	8
Russell E	4
Russell F	5
Russell R	3
Russell S	1
Schiaparelli A	17
Seleucus	
Seleucus A	19
Seleucus E	18
Struve	
Struve B	24
Struve C	15
Struve D	9
Struve F	16
Struve G	12
Struve H	10
Struve K	13
Struve L	21
Struve M	14
Vasco da Gama	
Vasco da Gama B	37
Vasco da Gama F	40
Voskresenskiy K	2

Dorsa Whiston
Aloha
43
44
1
2
3
Rima Krieger
45
46
Montes Agricola
VIII
Rima Aristarchus VII
4
47
Dorsum Niggli
Krieger
Rima Agricola
VI
48
5
Van Biesbroeck
49
Rimae Aristarchus
6
50
Toscanelli
7
Golgi
51
8
Mons Herodotus
9
Dorsa Burnet
Rupes Toscanelli
V
IV
Raman
10
III
Zinner
12
13
Väisälä
11
52
Freud
Vallis Schröteri
67
14
Prinz
53
Rima Aristarchus I
II
15
16
54
55
17
Aristarchus
Schiaparelli
Herodotus
18
21
56
19
20
22
23
25
24
OCEANUS
26
57
27
28
29
30
31
PROCELLARUM
32
33
Rima Marius
34
35
36
37
38
58
39
60
59
61
40
62
63
64
65
41
66
42

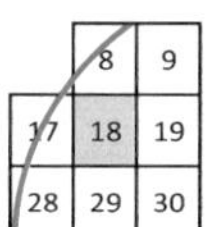

18

No.	Name	No.	Name	Name	No.	Name	No.
	Aloha	30	Aristarchus S	Aloha		Marius γ	64
	Montes Agricola	31	Marius N	Aristarchus		Marius δ	62
	Rima Krieger	32	Marius P	Aristarchus B	13	Marius ζ	65
	Krieger	33	Marius M	Aristarchus F	25	Marius θ	60
	Dorsa Whiston	34	Bessarion B	Aristarchus H	20	Marius μ	66
	Rima Agricola	35	Marius Q	Aristarchus N	21	Marius ϕ	61
	Dorsum Niggli	36	Marius B	Aristarchus S	30	Mons Herodotus	
	Rimae Aristarchus	37	Bessarion C	Aristarchus T	29	Montes Agricola	
	Rima Aristarchus I	38	Marius L	Aristarchus U	28	Oceanus Procellarum	
	Rima Aristarchus II	39	Bessarion H	Aristarchus Z	14	Prinz	
	Rima Aristarchus III	40	Bessarion G	Aristarchus δ	45	Raman	
	Rima Aristarchus IV	41	Marius C	Aristarchus μ	50	Rima Agricola	
	Rima Aristarchus V	42	Marius S	Bessarion B	34	Rima Aristarchus I	
	Rima Aristarchus VI	43	Herodotus ι	Bessarion C	37	Rima Aristarchus II	
	Rima Aristarchus VII	44	Wollaston α	Bessarion D	27	Rima Aristarchus III	
	Rima Aristarchus VIII	45	Aristarchus δ	Bessarion G	40	Rima Aristarchus IV	
	Van Biesbroeck	46	Herodotus κ	Bessarion H	39	Rima Aristarchus V	
	Toscanelli	47	Lichtenberg AA	Brayley L	26	Rima Aristarchus VI	
	Golgi	48	Herodotus γ	Dorsa Burnet		Rima Aristarchus VII	
	Mons Herodotus	49	Herodotus ν	Dorsa Whiston		Rima Aristarchus VIII	
	Rupes Toscanelli	50	Aristarchus μ	Dorsum Niggli		Rima Krieger	
	Raman	51	Herodotus η	Freud		Rima Marius	
	Zinner	52	Herodotus σ	Golgi		Rimae Aristarchus	
	Dorsa Burnet	53	Herodotus λ	Herodotus		Rupes Toscanelli	
	Vallis Schröteri	54	Herodotus δ	Herodotus A	24	Schiaparelli	
	Freud	55	Herodotus ρ	Herodotus B	19	Schiaparelli A	18
	Prinz	56	Herodotus τ	Herodotus C	23	Schiaparelli C	11
	Väisälä	57	Herodotus ω	Herodotus E	1	Schiaparelli E	9
	Aristarchus	58	Marius BC	Herodotus G	15	Seleucus A	22
	Herodotus	59	Marius BA	Herodotus H	10	Toscanelli	
	Schiaparelli	60	Marius θ	Herodotus K	16	Väisälä	
	Rima Marius	61	Marius ϕ	Herodotus L	12	Vallis Schröteri	
	Oceanus Procellarum	62	Marius δ	Herodotus N	17	Van Biesbroeck	
1	Herodotus E	63	Marius BB	Herodotus R	8	Wollaston N	5
2	Wollaston R	64	Marius γ	Herodotus S	7	Wollaston P	3
3	Wollaston P	65	Marius ζ	Herodotus T	6	Wollaston R	2
4	Lichtenberg A	66	Marius μ	Herodotus γ	48	Wollaston α	44
5	Wollaston N	67	Herodotus θ	Herodotus δ	54	Zinner	
6	Herodotus T			Herodotus η	51		
7	Herodotus S			Herodotus θ	67		
8	Herodotus R			Herodotus ι	43		
9	Schiaparelli E			Herodotus κ	46		
10	Herodotus H			Herodotus λ	53		
11	Schiaparelli C			Herodotus ν	49		
12	Herodotus L			Herodotus ρ	55		
13	Aristarchus B			Herodotus σ	52		
14	Aristarchus Z			Herodotus τ	56		
15	Herodotus G			Herodotus ω	57		
16	Herodotus K			Krieger			
17	Herodotus N			Lichtenberg A	4		
18	Schiaparelli A			Lichtenberg AA	47		
19	Herodotus B			Marius B	36		
20	Aristarchus H			Marius BA	59		
21	Aristarchus N			Marius BB	63		
22	Seleucus A			Marius BC	58		
23	Herodotus C			Marius C	41		
24	Herodotus A			Marius L	38		
25	Aristarchus F			Marius M	33		
26	Brayley L			Marius N	31		
27	Bessarion D			Marius P	32		
28	Aristarchus U			Marius Q	35		
29	Aristarchus T			Marius S	42		

35
Angström
36
Mons Delisle
Delisle
Rima Diophantus
Krieger
Rocco
Ruth
Van Biesbroeck
R.A. VI
1
2
Samir
Louise
Isabel
Walter
Fedorov
37
Dorsa Argand
Rimae Aristarchus
3
Rimae Prinz
38
R.A. V
R.A. IV
R.A. III
R.P. I
R.P. II
Montes Harbinger
Ivan
Artsimovich
4
Diophantus
5
Rima Artsimovich
39
40
Vera
41
Prinz
42
43
44
6
Dorsum Arduino
45
7
8
Courtney
Rima Zahia
Catena Yuri
Dorsum Thera
9
Euler
10
Mons Vinogradov
11
Rima Brayley
46
12
13
14
15
16
19
Brayley
47
17
18
Rima Euler
Ango
Jehan
Rima Wan-Yu
48
Rosa
20
Akis
Natasha
49
Catena Pierre
21
50
22
23
51
52
24
25
26
53
27
54
55
T. Mayer
30
31
29
28
32
33
56
57
Bessarion
59
58
60
61
62
63
34
Rima T. Mayer
64
65

8	9	10
18	19	20
29	30	31

19

No.	Name	No.	Name	Name	No.	Name	No.
	Angström	13	Brayley K	Akis		Harbringer μ	38
	Delisle	14	Brayley E	Ango		Isabel	
	Krieger	15	Brayley L	Angström		Ivan	
	Rocco	16	Brayley F	Aristarchus D	9	Jehan	
	Ruth	17	Euler L	Aristarchus N	10	Krieger	
	Van Biesbroeck	18	Euler F	Artsimovich		Krieger C	3
	Mons Delisle	19	Brayley B	Bessarion		Louise	
	Samir	20	Brayley D	Bessarion A	25	Mons Delisle	
	Louise	21	Bessarion D	Bessarion B	26	Mons Vinogradov	
	Isabel	22	T. Mayer K	Bessarion D	21	Montes Carpatus	
	Walter	23	T. Mayer W	Bessarion E	29	Montes Harbinger	
	Rima Diophantus	24	T. Mayer G	Bessarion G	33	Natasha	
	Dorsa Argand	25	Bessarion A	Bessarion H	28	Prinz	
	Fedorov	26	Bessarion B	Bessarion V	32	Rima Aristarchus III	
	Artsimovich	27	Bessarion W	Bessarion W	27	Rima Aristarchus IV	
	Diophantus	28	Bessarion H	Bessarion ζ	57	Rima Aristarchus V	
	Montes Harbinger	29	Bessarion E	Bessarion η	60	Rima Aristarchus VI	
	Rimae Prinz	30	T. Mayer B	Bessarion θ	56	Rima Artsimovich	
	Rima Prinz I	31	T. Mayer A	Bessarion λ	64	Rima Brayley	
	Rima Prinz II	32	Bessarion V	Bessarion ξ	61	Rima Diophantus	
	Rimae Aristarchus	33	Bessarion G	Brayley		Rima Euler	
	Rima Aristarchus III	34	T. Mayer P	Brayley B	19	Rima Prinz I	
	Rima Aristarchus IV	35	Wollaston α	Brayley C	12	Rima Prinz II	
	Rima Aristarchus V	36	Delisle ϵ	Brayley D	20	Rima T. Mayer	
	Rima Aristarchus VI	37	Harbringer δ	Brayley E	14	Rima Wan-Yu	
	Ivan	38	Harbringer μ	Brayley F	16	Rima Zahia	
	Vera	39	Harbringer ζ	Brayley G	8	Rimae Aristarchus	
	Prinz	40	Harbringer β	Brayley K	13	Rimae Prinz	
	Rima Artsimovich	41	Harbringer θ	Brayley L	15	Rocco	
	Dorsum Arduino	42	Harbringer γ	Brayley S	6	Rosa	
	Courtney	43	Harbringer α	Brayley α	46	Ruth	
	Rima Zahia	44	Brayley σ	Brayley π	45	Samir	
	Dorsum Thera	45	Brayley π	Brayley σ	44	T. Mayer	
	Catena Yuri	46	Brayley α	Catena Pierre		T. Mayer A	31
	Euler	47	Euler π	Catena Yuri		T. Mayer B	30
	Mons Vinogradov	48	Euler δ	Courtney		T. Mayer G	24
	Rima Brayley	49	Euler ν	Delisle		T. Mayer K	22
	Brayley	50	Euler γ	Delisle K	1	T. Mayer P	34
	Rima Euler	51	T. Mayer ρ	Delisle ϵ	36	T. Mayer W	23
	Ango	52	T. Mayer β	Diophantus		T. Mayer α	65
	Jehan	53	T. Mayer λ	Diophantus B	2	T. Mayer β	52
	Rosa	54	T. Mayer σ	Diophantus C	4	T. Mayer δ	62
	Akis	55	T. Mayer ξ	Diophantus D	5	T. Mayer η	59
	Rima Wan-Yu	56	Bessarion θ	Dorsa Argand		T. Mayer λ	53
	Catena Pierre	57	Bessarion ζ	Dorsum Arduino		T. Mayer ξ	55
	Natasha	58	T. Mayer π	Dorsum Thera		T. Mayer π	58
	Bessarion	59	T. Mayer η	Euler		T. Mayer ρ	51
	Rima T. Mayer	60	Bessarion η	Euler E	7	T. Mayer σ	54
	T. Mayer	61	Bessarion ξ	Euler F	18	T. Mayer ω	63
	Montes Carpatus	62	T. Mayer δ	Euler J	11	Van Biesbroeck	
1	Delisle K	63	T. Mayer ω	Euler L	17	Vera	
2	Diophantus B	64	Bessarion λ	Euler γ	50	Walter	
3	Krieger C	65	T. Mayer α	Euler δ	48	Wollaston α	35
4	Diophantus C			Euler ν	49		
5	Diophantus D			Euler π	47		
6	Brayley S			Fedorov			
7	Euler E			Harbringer α	43		
8	Brayley G			Harbringer β	40		
9	Aristarchus D			Harbringer γ	42		
10	Aristarchus N			Harbringer δ	37		
11	Euler J			Harbringer ζ	39		
12	Brayley C			Harbringer θ	41		

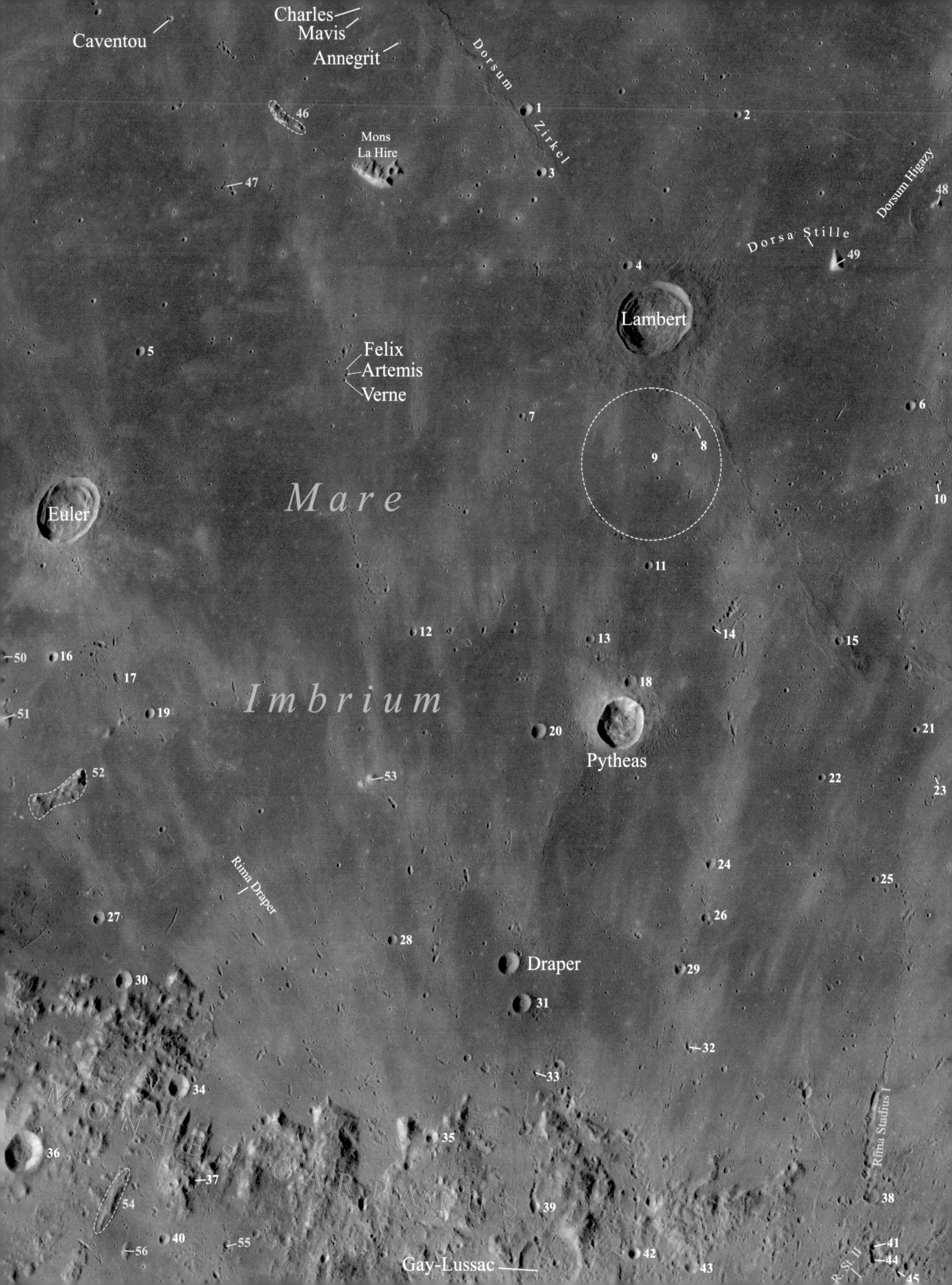
Charles
Mavis
Caventou
Annegrit
Dorsum Zirkel
Mons La Hire
Dorsum Higazy
Dorsa Stille
Lambert
Felix
Artemis
Verne
Euler
Mare Imbrium
Pytheas
Rima Draper
Draper
Montes Carpatus
Rima Stadius I
R. St. II
Gay-Lussac
1
2
3
4
5
6
7
8
9
10
11
12
13
14
15
16
17
18
19
20
21
22
23
24
25
26
27
28
29
30
31
32
33
34
35
36
37
38
39
40
41
42
43
44
45
46
47
48
49
50
51
52
53
54
55
56

No.	Name
	Charles
	Mavis
	Caventou
	Annegrlt
	Dorsum Zirkel
	Mons La Hire
	Dorsum Higazy
	Dorsa Stille
	Lambert
	Felix
	Artemis
	Verne
	Mare Imbrium
	Euler
	Pytheas
	Rima Draper
	Draper
	Montes Carpatus
	Rima Stadius I
	Rima Stadius II
	Gay-Lussac
1	La Hire A
2	Lambert T
3	La Hire B
4	Lambert A
5	Euler H
6	Timocharis E
7	Lambert W
8	Lambert B
9	Lambert R
10	Timocharis H
11	Pytheas N
12	Pytheas W
13	Pytheas J
14	Pytheas U
15	Pytheas G
16	Euler L
17	Euler F
18	Pytheas D
19	Euler G
20	Pytheas A
21	Pytheas H
22	Pytheas M
23	Pytheas K
24	Pytheas C
25	Pytheas L
26	Pytheas E
27	T. Mayer K
28	Draper A
29	Pytheas B
30	T. Mayer G
31	Draper C
32	Pytheas F
33	Gay-Lussac B
34	T. Mayer E
35	Gay-Lussac C
36	T. Mayer A
37	T. Mayer M
38	Stadius M
39	Gay-Lussac D
40	T. Mayer Z
41	Stadius W
42	Gay-Lussac F
43	Gay-Lussac G
44	Stadius U
45	Stadius J
46	Prom La Hire α
47	Prom La Hire C
48	Lambert δ
49	Lambert γ
50	Euler π
51	Euler δ
52	Euler γ
53	Pytheas β
54	T. Mayer η
55	T. Mayer J
56	T. Mayer ω

Name	No.
Annegrit	
Artemis	
Caventou	
Charles	
Dorsa Stille	
Dorsum Higazy	
Dorsum Zirkel	
Draper	
Draper A	28
Draper C	31
Euler	
Euler F	17
Euler G	19
Euler H	5
Euler L	16
Euler γ	52
Euler δ	51
Euler π	50
Felix	
Gay-Lussac	
Gay-Lussac B	33
Gay-Lussac C	35
Gay-Lussac D	39
Gay-Lussac F	42
Gay-Lussac G	43
La Hire A	1
La Hire B	3
Lambert	
Lambert A	4
Lambert B	8
Lambert R	9
Lambert T	2
Lambert W	7
Lambert γ	49
Lambert δ	48
Mare Imbrium	
Mavis	
Mons La Hire	
Montes Carpatus	
Prom La Hire C	47
Prom La Hire α	46
Pytheas	
Pytheas A	20
Pytheas B	29
Pytheas C	24
Pytheas D	18
Pytheas E	26
Pytheas F	32
Pytheas G	15
Pytheas H	21
Pytheas J	13
Pytheas K	23
Pytheas L	25
Pytheas M	22
Pytheas N	11
Pytheas U	14
Pytheas W	12
Pytheas β	53
Rima Draper	
Rima Stadius I	
Rima Stadius II	
Stadius J	45
Stadius M	38
Stadius U	44
Stadius W	41
T. Mayer A	36
T. Mayer E	34
T. Mayer G	30
T. Mayer J	55
T. Mayer K	27
T. Mayer M	37
T. Mayer Z	40
T. Mayer η	54
T. Mayer ω	56
Timocharis E	6
Timocharis H	10
Verne	

1
Dorsum Grabau
Sampson
Catena Timocharis
2
37
38
Dorsum Higazy
3
4
Feuillée
Bancroft
II
5
39
Rimae Archimedes
Beer
I
Timocharis
6
7
40
8
9
10
Heinrich
Macmillan
11
12
13
36
Pupin
Mare
Imbrium
14
15
Wallace
16
17
18
19
20
21
22
23
25
24
26
41
Mons Wolff
27
28
29
Montes Apenninus
42
43
30
Eratosthenes
Sinus Aestuum
31
33
32
34
35

10	11	12
20	21	22
31	32	33

21

No.	Name	No.	Name	Name	No.	Name	No.
	[illegible]	[illegible]	[illegible]	[illegible]	[illegible]	Wolff A	[illegible]
	Sampson	43	Marco Polo L	Archimedes AB	37	Wolff B	28
	Catena Timocharis			Archimedes E	8		
	Dorsum Higazy			Archimedes G	2		
	Feuillée			Archimedes H	12		
	Bancroft			Archimedes R	6		
	Rima Archimedes II			Archimedes W	13		
	Rimae Archimedes			Archimedes Y	1		
	Rima Archimedes I			Bancroft			
	Beer			Beer			
	Timocharis			Beer A	5		
	Montes Archimedes			Beer B	7		
	Heinrich			Beer E	4		
	Pupin			Catena Timocharis			
	Macmillan			Dorsum Grabau			
	Mare Imbrium			Dorsum Higazy			
	Wallace			Eratosthenes			
	Mons Wolff			Eratosthenes A	21		
	Montes Apenninus			Eratosthenes B	19		
	Eratosthenes			Eratosthenes C	27		
	Sinus Aestuum			Eratosthenes D	26		
1	Archimedes Y			Eratosthenes E	22		
2	Archimedes G			Eratosthenes F	25		
3	Timocharis B			Eratosthenes G	42		
4	Beer E			Eratosthenes M	32		
5	Beer A			Eratosthenes Z	35		
6	Archimedes R			Feuillée			
7	Beer B			Heinrich			
8	Archimedes E			Lambert δ	39		
9	Timocharis C			Macmillan			
10	Timocharis E			Marco Polo C	31		
11	Timocharis D			Marco Polo E	41		
12	Archimedes H			Marco Polo L	43		
13	Archimedes W			Mare Imbrium			
14	Wallace H			Mons Wolff			
15	Pytheas H			Montes Apenninus			
16	Pytheas K			Montes Archimedes			
17	Wallace K			Pupin			
18	Wallace A			Pytheas H	15		
19	Eratosthenes B			Pytheas K	16		
20	Pytheas L			Pytheas L	20		
21	Eratosthenes A			Rima Archimedes I			
22	Eratosthenes E			Rima Archimedes II			
23	Wallace D			Rimae Archimedes			
24	Wallace C			Sampson			
25	Eratosthenes F			Sinus Aestuum			
26	Eratosthenes D			Stadius J	34		
27	Eratosthenes C			Stadius U	33		
28	Wolff B			Stadius W	30		
29	Wolff A			Timocharis			
30	Stadius W			Timocharis AA	40		
31	Marco Polo C			Timocharis B	3		
32	Eratosthenes M			Timocharis C	9		
33	Stadius U			Timocharis D	11		
34	Stadius J			Timocharis E	10		
35	Eratosthenes Z			Timocharis H	36		
36	Timocharis H			Wallace			
37	Archimedes AB			Wallace A	18		
38	Archimedes AA			Wallace C	24		
39	Lambert δ			Wallace D	23		
40	Timocharis AA			Wallace H	14		
41	Marco Polo E			Wallace K	17		

Archimedes
Bancroft
Spurr
Palus Putredinis
Promontorium Fresnel
Rimae Fresnel
Santos-Dumont
Mons Hadley
Detail
Julienne
St. George
Rima Hadley
Mons Hadley Delta
Carlos
Taizo
Bela
Jomo
Rimae Archimedes
Montes Archimedes
Ian
Kathleen
Rima Vladimir
Michael
Ann
Patricia
Rima Bradley
Aratus
Mons Bradley
Conon
Galen
Huxley
Mons Ampere
Mons Huygens
Montes Apenninus
Rima Conon
Lacus Felicitatis
Sinus Fidei
Yangel'
Rima Yangel'
Marco Polo
Mare Vaporum

22

11	12	13
21	22	23
32	33	34

No.	Name	No.	Name	Name	No.	Name	No.
	Archimedes	15	Huygens A	Ann		Marco Polo S	23
	Promontorium Fresnel	16	Conon A	Annegrit		Marco Polo T	33
	Rimae Fresnel	17	Wallace A	Aratus		Marco Polo γ	52
	Rima Fresnel I	18	Conon W	Aratus B	9	Mare Vaporum	
	Rima Fresnel II	19	Marco Polo K	Aratus θ	43	Michael	
	Rima Fresnel III	20	Wallace D	Archimedes		Mons Ampère	
	Bancroft	21	Marco Polo H	Archimedes L	7	Mons Bradley	
	Spurr	22	Marco Polo J	Archimedes M	4	Mons Hadley	
	Santos-Dumont	23	Marco Polo S	Archimedes N	8	Mons Hadley Delta	
	Palus Putredinis	24	Marco Polo M	Archimedes P	5	Mons Huygens	
	Mons Hadley	25	Marco Polo B	Archimedes Q	2	Montes Apenninus	
	Mons Hadley Delta	26	Marco Polo P	Archimedes S	1	Montes Archimedes	
	Rimae Archimedes	27	Marco Polo G	Archimedes Z	3	Palus Putredinis	
	Rima Archimedes III	28	Marco Polo F	Archimedes γ	41	Patricia	
	Rima Archimedes IV	29	Marco Polo L	Archimedes δ	42	Promontorium Fresnel	
	Rima Archimedes V	30	Marco Polo D	Archimedes μ	38	Rima Archimedes III	
	Rima Archimedes VI	31	Marco Polo A	Archimedes φ	39	Rima Archimedes IV	
	Julienne	32	Marco Polo C	Autolycus B	34	Rima Archimedes V	
	Montes Archimedes	33	Marco Polo T	Autolycus α	35	Rima Archimedes VI	
	St. George	34	Autolycus B	Autolycus β	37	Rima Bradley	
	Rima Hadley	35	Autolycus α	Autolycus γ	36	Rima Conon	
	Ian	36	Autolycus γ	Bancroft		Rima Fresnel I	
	Kathleen	37	Autolycus β	Béla		Rima Fresnel II	
	Rima Vladimir	38	Archimedes μ	Bradley H	11	Rima Fresnel III	
	Ann	39	Archimedes φ	Bradley K	10	Rima Hadley	
	Annegrit	40	Hadley δ	Bradley φ	46	Rima Vladimir	
	Michael	41	Archimedes γ	Carlos		Rima Yangel'	
	Patricia	42	Archimedes δ	Conon		Rimae Archimedes	
	Carlos	43	Aratus θ	Conon A	16	Rimae Fresnel	
	Taizo	44	Huygens M	Conon W	18	Santos-Dumont	
	Béla	45	Conon λ	Conon Y	12	Sinus Fidei	
	Jomo	46	Bradley φ	Conon Z	50	Spurr	
	Rima Bradley	47	Huygens β	Conon α	48	St. George	
	Aratus	48	Conon α	Conon β	49	Sulpicius Gallus H	14
	Conon	49	Conon β	Conon λ	45	Taizo	
	Mons Bradley	50	Conon Z	Conon φ	51	Wallace A	17
	Galen	51	Conon φ	Galen		Wallace D	20
	Huxley	52	Marco Polo γ	Hadley C	6	Wallace T	13
	Mons Huygens	53	Manilius β	Hadley δ	40	Yangel'	
	Mons Ampère			Huxley			
	Montes Apenninus			Huygens A	15		
	Rima Conon			Huygens M	44		
	Lacus Felicitatis			Huygens β	47		
	Sinus Fidei			Ian			
	Yangel'			Jomo			
	Rima Yangel'			Julienne			
	Marco Polo			Kathleen			
	Mare Vaporum			Lacus Felicitatis			
1	Archimedes S			Manilius β	53		
2	Archimedes Q			Marco Polo			
3	Archimedes Z			Marco Polo A	31		
4	Archimedes M			Marco Polo B	25		
5	Archimedes P			Marco Polo C	32		
6	Hadley C			Marco Polo D	30		
7	Archimedes L			Marco Polo F	28		
8	Archimedes N			Marco Polo G	27		
9	Aratus B			Marco Polo H	21		
10	Bradley K			Marco Polo J	22		
11	Bradley H			Marco Polo K	19		
12	Conon Y			Marco Polo L	29		
13	Wallace T			Marco Polo M	24		
14	Sulpicius Gallus H			Marco Polo P	26		

22
23
24 25 1 2
Mare
Linné
26
27 28
29 30
Banting
Dorsum Von Cotta
Detail 4
31
Serenitatis
Monte Apenninus
Joy
32
33
Dorsum Gast
3
4
5
Hornsby
34
35
36
Dorsum Buckland
6
37
Rimae Sulpicius Gallus
Bessel
Rima Sulpicius Gallus
II
38
39
7
8
III
9
10
Rima Sulpicius Gallus I
11
Sulpicius Gallus
Bobillier
Lacus Odii
40
41
Detail 3
Dorsa Sorby
Monte Haemus
Lacus
Felicitatis
21
42
12
43
13
44
Bowen
Rimae Menelaus
Lacus Doloris
14
45
15
Lacus Gaudii
16
46
Menelaus
47
Daubrée
17
50
Lacus Hiemalis
48
49
18
Mare
51
Manilius
Lacus Lenitatis
19
Vaporum
20
52

12	13	14
22	23	24
33	34	35

23

No.	Name
	Mare Serenitatis
	Linné
	Banting
	Dorsum von Cotta
	Joy
	Dorsum Gast
	Montes Apenninus
	Hornsby
	Bessel
	Dorsum Buckland
	Rimae Sulpicius Gallus
	Rima Sulpicius Gallus I
	Rima Sulpicius Gallus II
	Rima Sulpicius Gallus III
	Sulpicius Gallus
	Bobillier
	Dorsa Sorby
	Lacus Odii
	Lacus Felicitatis
	Montes Haemus
	Rimae Menelaus
	Bowen
	Lacus Doloris
	Lacus Gaudii
	Menelaus
	Daubrée
	Lacus Hiemalis
	Lacus Lenitatis
	Manilius
	Mare Vaporum
1	Linné A
2	Linné D
3	Aratus D
4	Aratus B
5	Aratus C
6	Sulpicius Gallus A
7	Bessel F
8	Bessel G
9	Sulpicius Gallus H
10	Sulpicius Gallus M
11	Sulpicius Gallus G
12	Sulpicius Gallus B
13	Manilius H
14	Menelaus A
15	Manilius B
16	Manilius Z
17	Manilius G
18	Menelaus C
19	Manilius X
20	Manilius U
21	Manilius E
22	Fresnel ψ
23	Linné AE
24	Linné AD
25	Linné ED
26	Linné ED
27	Linné AB
28	Linné EB
29	Linné AC
30	Linné EC
31	Linné EA
32	Linné EF
33	Aratus CA
34	Aratus CC
35	Bessel FA
36	Bessel FC
37	Bessel FB
38	Bessel GB
39	Bessel GA
40	Manilius EA
41	Sulpicius Gallus BA
42	Sulpicius Gallus α
43	Sulpicius Gallus BB
44	Menelaus ζ
45	Manilius FA
46	Menelaus AB
47	Manilius α
48	Manilius GA
49	Manilius β
50	Menelaus R
51	Manilius N
52	Menelaus α

Name	No.
Aratus B	4
Aratus C	5
Aratus CA	33
Aratus CC	34
Aratus D	3
Banting	
Bessel F	7
Bessel FA	35
Bessel FB	37
Bessel FC	36
Bessel G	8
Bessel GA	39
Bessel GB	38
Bobillier	
Bowen	
Daubrée	
Dorsa Sorby	
Dorsum Buckland	
Dorsum Gast	
Dorsum von Cotta	
Fresnel ψ	22
Hornsby	
Joy	
Lacus Doloris	
Lacus Felicitatis	
Lacus Gaudii	
Lacus Hiemalis	
Lacus Lenitatis	
Lacus Odii	
Linné	
Linné A	1
Linné AB	27
Linné AC	29
Linné AD	24
Linné AE	23
Linné D	2
Linné EA	31
Linné EB	28
Linné EC	30
Linné ED	25
Linné ED	26
Linné EF	32
Manilius	
Manilius B	15
Manilius E	21
Manilius EA	40
Manilius FA	45
Manilius G	17
Manilius GA	48
Manilius H	13
Manilius N	51
Manilius U	20
Manilius X	19
Manilius Z	16
Manilius α	47
Manilius β	49
Mare Serenitatis	
Mare Vaporum	
Menelaus	
Menelaus A	14
Menelaus AB	46
Menelaus C	18
Menelaus R	50
Menelaus α	52
Menelaus ζ	44
Montes Apenninus	
Montes Haemus	
Rima Sulpicius Gallus I	
Rima Sulpicius Gallus II	
Rima Sulpicius Gallus III	
Rimae Menelaus	
Rimae Sulpicius Gallus	
Sulpicius Gallus	
Sulpicius Gallus A	6
Sulpicius Gallus B	12
Sulpicius Gallus BA	41
Sulpicius Gallus BB	43
Sulpicius Gallus G	11
Sulpicius Gallus H	9
Sulpicius Gallus M	10
Sulpicius Gallus α	42

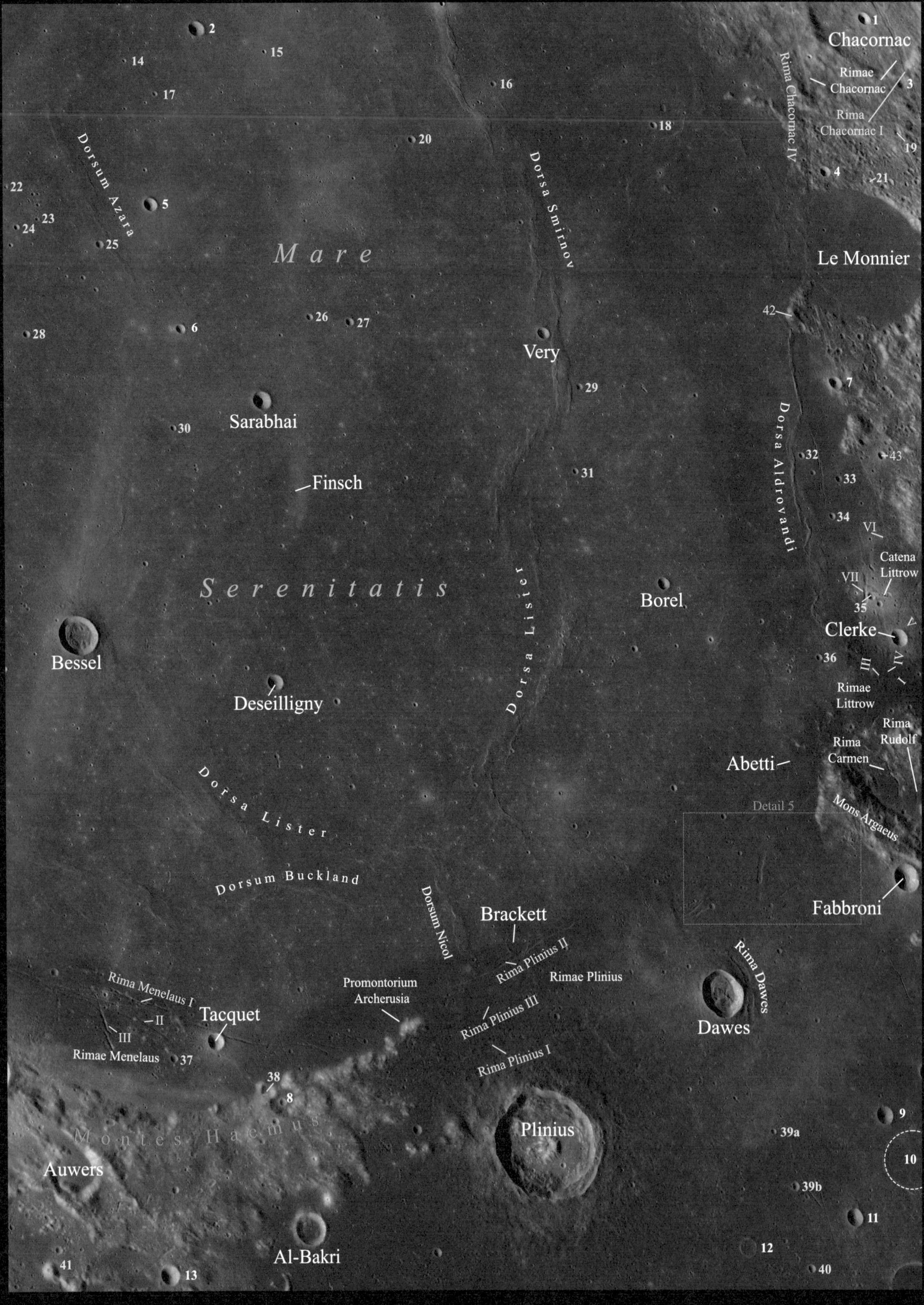

1
Chacornac
2
15
14
Rimae Chacornac
3
Rima Chacornac IV
16
17
Rima Chacornac I
19
18
20
Dorsum Azara
Dorsa Smirnov
4
21
22
5
23
24
25
Mare
Le Monnier
42
26
27
6
28
Very
29
7
Sarabhai
30
Dorsa Aldrovandi
32
43
33
31
Finsch
34
VI
Catena Littrow
VII
Serenitatis
Borel
35
Bessel
Clerke
V
Dorsa Lister
36
IV
III
I
Rimae Littrow
Deseilligny
Rima Rudolf
Rima Carmen
Abetti
Dorsa Lister
Detail 5
Mons Argaeus
Dorsum Buckland
Dorsum Nicol
Fabbroni
Brackett
Rima Plinius II
Rima Dawes
Rimae Plinius
Rima Menelaus I
Promontorium Archerusia
II
Tacquet
Rima Plinius III
Dawes
III
Rimae Menelaus
37
Rima Plinius I
38
8
Montes Haemus
Plinius
9
39a
10
Auwers
39b
11
Al-Bakri
12
41
13
40

13	14	15
23	24	25
34	35	36

24

No.	Name
	Chacornac
	Rima Chacornac I
	Rima Chacornac IV
	Dorsa Smirnov
	Dorsum Azara
	Mare Serenitatis
	Le Monnier
	Very
	Dorsa Aldrovandi
	Sarabhai
	Finsch
	Borel
	Catena Littrow
	Bessel
	Clerke
	Dorsa Lister
	Rimae Littrow
	Rima Littrow I
	Rima Littrow III
	Rima Littrow IV
	Rima Littrow V
	Rima Littrow VI
	Rima Littrow VII
	Deseilligny
	Rima Rudolf
	Rima Carmen
	Abetti
	Mons Argaeus
	Dorsum Buckland
	Fabbroni
	Dorsum Nicol
	Brackett
	Rimae Plinius
	Rima Plinius I
	Rima Plinius II
	Rima Plinius III
	Rima Dawes
	Dawes
	Promontorium Archerusia
	Rima Menelaus I
	Rima Menelaus II
	Rima Menelaus III
	Tacquet
	Rimae Menelaus
	Montes Haemus
	Plinius
	Auwers
	Al-Bakri
1	Chacornac A
2	Posidonius N
3	Chacornac B
4	Le Monnier K
5	Bessel D
6	Bessel H
7	Le Monnier H
8	Tacquet B
9	Jansen D
10	Jansen R
11	Jansen E
12	Plinius B
13	Auwers A
14	Posidonius NB
15	Posidonius NC
16	Posidonius NE
17	Posidonius NA
18	Le Monnier KA
19	Chacornac BA
20	Posidonius ND
21	Le Monnier KB
22	Linné ED
23	Linné EB
24	Linné EC
25	Bessel DA
26	Bessel AC
27	Bessel AB
28	Linné EA
29	Le Monnier BM
30	Bessel AD
31	Le Monnier BL
32	Le Monnier LC
33	Le Monnier LB
34	Le Monnier LA
35	Littrow BA
36	Littrow BB
37	Tacquet BB
38	Tacquet BA
40	Jansen EA
41	Menelaus α
42	Le Monnier α
43	Le Monnier L
39a	Jansen EB
39b	Jansen EB

Name	No.
Abetti	
Al-Bakri	
Auwers	
Auwers A	13
Bessel	
Bessel AB	27
Bessel AC	26
Bessel AD	30
Bessel D	5
Bessel DA	25
Bessel H	6
Borel	
Brackett	
Catena Littrow	
Chacornac	
Chacornac A	1
Chacornac B	3
Chacornac BA	19
Clerke	
Dawes	
Deseilligny	
Dorsa Aldrovandi	
Dorsa Lister	
Dorsa Smirnov	
Dorsum Azara	
Dorsum Buckland	
Dorsum Nicol	
Fabbroni	
Finsch	
Jansen D	9
Jansen E	11
Jansen EA	40
Jansen EB	39a
Jansen EB	39b
Jansen R	10
Le Monnier	
Le Monnier BL	31
Le Monnier BM	29
Le Monnier H	7
Le Monnier K	4
Le Monnier KA	18
Le Monnier KB	21
Le Monnier L	43
Le Monnier LA	34
Le Monnier LB	33
Le Monnier LC	32
Le Monnier α	42
Linné EA	28
Linné EB	23
Linné EC	24
Linné ED	22
Littrow BA	35
Littrow BB	36
Menelaus α	41
Mons Argaeus	
Montes Haemus	
Plinius	
Plinius B	12
Posidonius N	2
Posidonius NA	17
Posidonius NB	14
Posidonius NC	15
Posidonius ND	20
Posidonius NE	16
Promontorium Archerusia	
Rima Carmen	
Rima Chacornac I	
Rima Chacornac IV	
Rima Dawes	
Rima Littrow I	
Rima Littrow III	
Rima Littrow IV	
Rima Littrow V	
Rima Littrow VI	
Rima Littrow VII	
Rima Menelaus I	
Rima Menelaus II	
Rima Menelaus III	
Rima Plinius I	
Rima Plinius II	
Rima Plinius III	
Rima Rudolf	
Rimae Littrow	
Rimae Menelaus	
Rimae Plinius	
Sarabhai	
Tacquet	
Tacquet B	8
Tacquet BA	38
Tacquet BB	37
Very	

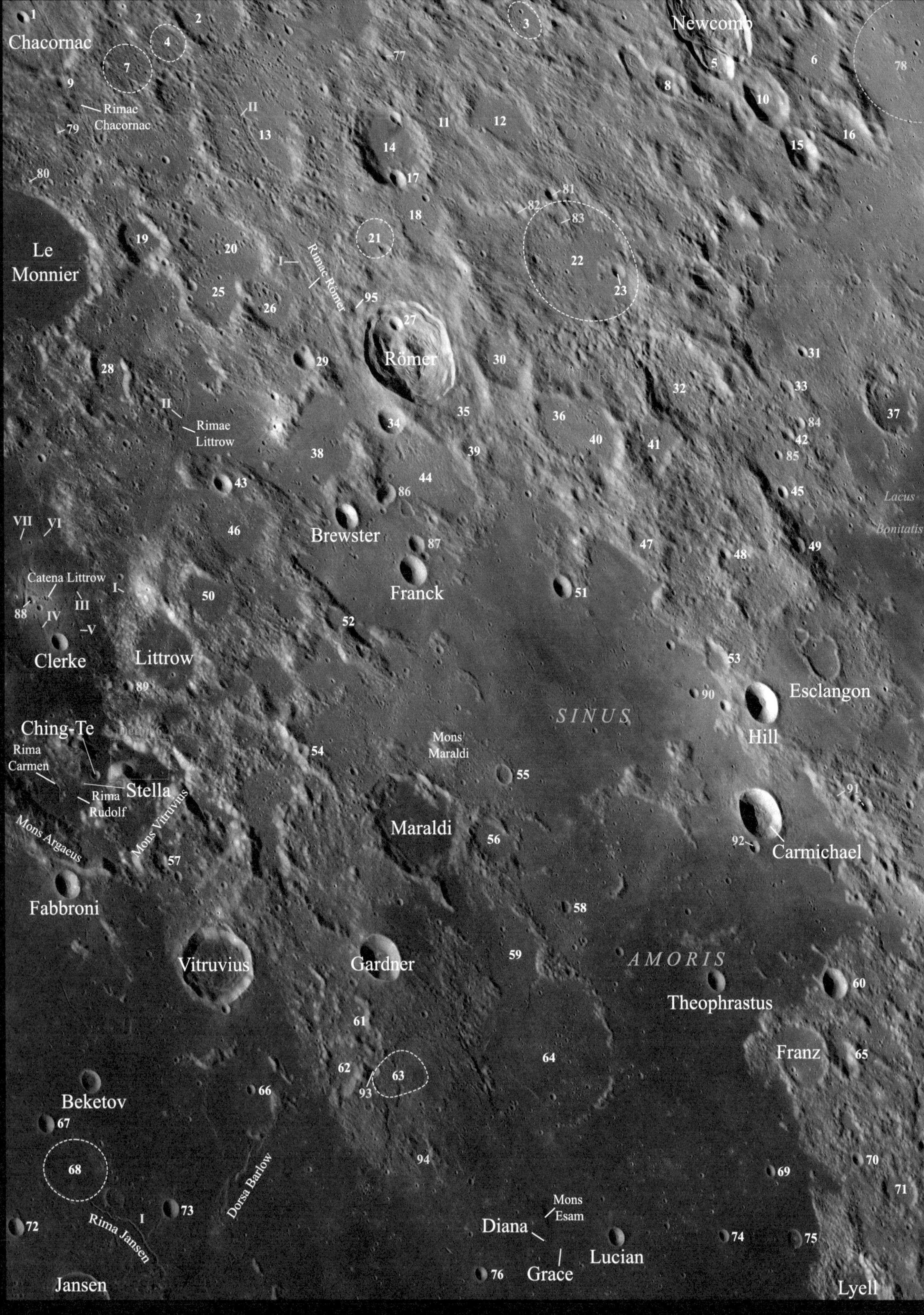

1
Chacornac
2
3
Newcomb
4
7
5
6
78
77
8
10
9
Rimae Chacornac
II
13
11
12
14
15
16
79
80
17
81
82
83
18
Le Monnier
19
20
21
I
Rimae Römer
22
23
25
26
95
27
Römer
29
28
30
31
32
33
37
II
Rimae Littrow
35
36
34
84
42
40
41
38
39
85
44
43
86
45
Lacus Bonitatis
VII
VI
46
Brewster
87
47
48
49
Catena Littrow
I
50
Franck
51
88
III
IV
V
52
Clerke
Littrow
53
Esclangon
89
90
Hill
SINUS
Ching-Te
Mons Maraldi
54
Rima Carmen
55
Stella
Rima Rudolf
91
Mons Argaeus
Mons Vitruvius
Maraldi
56
92
Carmichael
57
Fabbroni
58
59
AMORIS
Vitruvius
Gardner
Theophrastus
60
61
Franz
65
64
62
63
93
66
Beketov
67
94
70
68
69
71
Dorsa Barlow
Mons Esam
73
72
I
Rima Jansen
Diana
74
75
Lucian
Grace
76
Jansen
Lyell

14	15	16
24	25	26
35	36	37

25

No.	Name	No.	Name	Name	No.	Name	No.
	Chacornac	25	Le Monnier U	Beketov		Maraldi R	54
	Newcomb	26	Le Monnier V	Brewster		Mons Argaeus	
	Rimae Chacornac	27	Römer Y	Carmichael		Mons Esam	
	Montes Taurus	28	Le Monnier T	Catena Littrow		Mons Maraldi	
	Le Monnier	29	Römer M	Chacornac		Mons Vitruvius	
	Rimae Römer	30	Römer N	Chacornac A	1	Montes Taurus	
	Rima Römer II	31	Macrobius V	Chacornac B	9	Newcomb	
	Rima Römer I	32	Macrobius M	Chacornac BA	79	Newcomb A	5
	Römer	33	Macrobius U	Chacornac E	4	Newcomb B	16
	Rimae Littrow	34	Römer D	Chacornac F	7	Newcomb C	6
	Rima Littrow I	35	Römer S	Ching-Te		Newcomb G	15
	Rima Littrow II	36	Römer V	Clerke		Newcomb H	8
	Rima Littrow III	37	Macrobius W	Diana		Newcomb J	10
	Rima Littrow IV	38	Römer R	Dorsa Barlow		Newcomb P	78
	Rima Littrow V	39	Römer Z	Esclangon		Proclus D	60
	Rima Littrow VI	40	Römer U	Fabbroni		Proclus E	65
	Rima Littrow VII	41	Römer X	Franck		Rima Carmen	
	Brewster	42	Macrobius Z	Franz		Rima Jansen	
	Lacus Bonitatis	43	Littrow D	G. Bond B	2	Rima Littrow I	
	Franck	44	Römer T	G. Bond C	13	Rima Littrow II	
	Catena Littrow	45	Macrobius Y	Gardner		Rima Littrow III	
	Clerke	46	Littrow P	Grace		Rima Littrow IV	
	Littrow	47	Macrobius P	Hill		Rima Littrow V	
	Esclangon	48	Macrobius N	Jansen		Rima Littrow VI	
	Hill	49	Macrobius X	Jansen D	67	Rima Littrow VII	
	Mons Maraldi	50	Littrow A	Jansen E	72	Rima Römer I	
	Ching-Te	51	Römer J	Jansen L	73	Rima Römer II	
	Stella	52	Littrow F	Jansen R	68	Rima Rudolf	
	Rima Carmen	53	Macrobius K	Kirchhoff G	3	Rimae Chacornac	
	Rima Rudolf	54	Maraldi R	Lacus Bonitatis		Rimae Littrow	
	Mons Vitruvius	55	Maraldi A	Le Monnier		Rimae Römer	
	Maraldi	56	Maraldi F	Le Monnier A	19	Römer	
	Carmichael	57	Vitruvius L	Le Monnier KB	80	Römer A	14
	Mons Argaeus	58	Maraldi N	Le Monnier S	20	Römer B	11
	Fabbroni	59	Maraldi E	Le Monnier T	28	Römer BA	77
	Sinus Amoris	60	Proclus D	Le Monnier U	25	Römer C	17
	Vitruvius	61	Vitruvius T	Le Monnier V	26	Römer D	34
	Gardner	62	Vitruvius B	Littrow		Römer E	12
	Theophrastus	63	Vitruvius H	Littrow A	50	Römer F	18
	Franz	64	Maraldi D	Littrow BA	88	Römer G	21
	Beketov	65	Proclus E	Littrow BC	89	Römer H	95
	Dorsa Barlow	66	Vitruvius M	Littrow D	43	Römer J	51
	Mons Esam	67	Jansen D	Littrow F	52	Römer KA	87
	Rima Jansen	68	Jansen R	Littrow P	46	Römer M	29
	Lucian	69	Lyell C	Lucian		Römer N	30
	Jansen	70	Lyell K	Lyell		Römer P	22
	Lyell	71	Lyell D	Lyell A	75	Römer PA	83
	Diana	72	Jansen E	Lyell B	74	Römer PB	81
	Grace	73	Jansen L	Lyell C	69	Römer PC	82
1	Chacornac A	74	Lyell B	Lyell D	71	Römer R	38
2	G. Bond B	75	Lyell A	Lyell K	70	Römer S	35
3	Kirchhoff G	76	Vitruvius G	Macrobius AA	91	Römer T	44
4	Chacornac E	77	Römer BA	Macrobius AB	92	Römer TA	86
5	Newcomb A	78	Newcomb P	Macrobius BA	90	Römer U	40
6	Newcomb C	79	Chacornac BA	Macrobius K	53	Römer V	36
7	Chacornac F	80	Le Monnier KB	Macrobius M	32	Römer W	23
8	Newcomb H	81	Römer PB	Macrobius N	48	Römer X	41
9	Chacornac B	82	Römer PC	Macrobius P	47	Römer Y	27
10	Newcomb J	83	Römer PA	Macrobius U	33	Römer Z	39
11	Römer B	84	Macrobius ZA	Macrobius V	31	Sinus Amoris	
12	Römer E	85	Macrobius ZB	Macrobius W	37	Stella	
13	G. Bond C	86	Römer TA	Macrobius X	49	Theophrastus	
14	Römer A	87	Römer KA	Macrobius Y	45	Vitruvius	
15	Newcomb G	88	Littrow BA	Macrobius Z	42	Vitruvius B	62
16	Newcomb B	89	Littrow BC	Macrobius ZA	84	Vitruvius C	94
17	Römer C	90	Macrobius BA	Macrobius ZB	85	Vitruvius G	76
18	Römer F	91	Macrobius AA	Maraldi		Vitruvius H	63
19	Le Monnier A	92	Macrobius AB	Maraldi A	55	Vitruvius K	93
20	Le Monnier S	93	Vitruvius K	Maraldi D	64	Vitruvius L	57
21	Römer G	94	Vitruvius C	Maraldi E	59	Vitruvius M	66
22	Römer P	95	Römer H	Maraldi F	56	Vitruvius T	61
23	Römer W			Maraldi N	58		

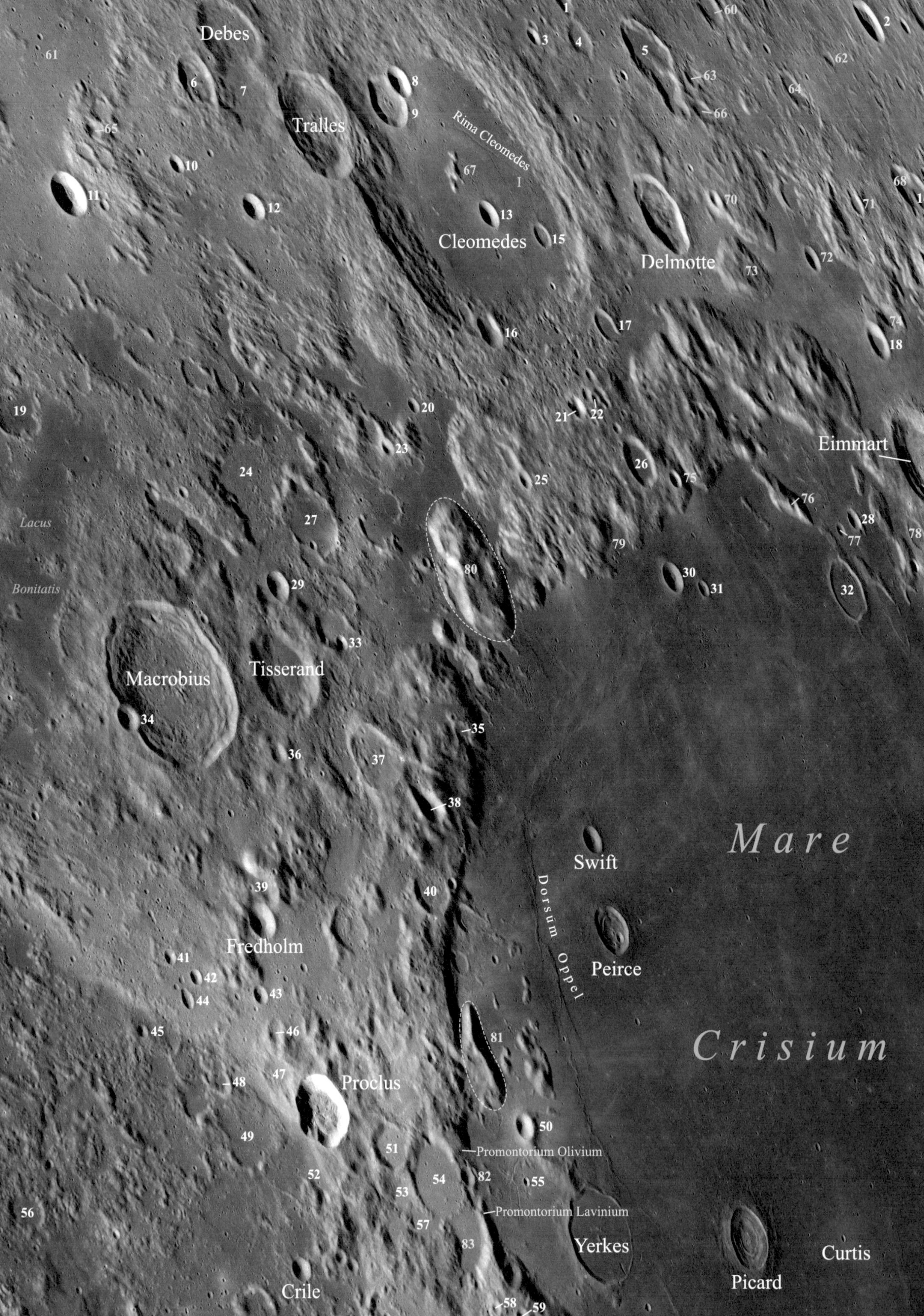
Debes
Tralles
Rima Cleomedes
I
Cleomedes
Delmotte
Eimmart
Lacus
Bonitatis
Macrobius
Tisserand
Swift
Mare
Dorsum Oppel
Peirce
Fredholm
Crisium
Proclus
Promontorium Olivium
Promontorium Lavinium
Yerkes
Curtis
Picard
Crile
1
2
3
4
5
6
7
8
9
10
11
12
13
14
15
16
17
18
19
20
21
22
23
24
25
26
27
28
29
30
31
32
33
34
35
36
37
38
39
40
41
42
43
44
45
46
47
48
49
50
51
52
53
54
55
56
57
58
59
60
61
62
63
64
65
66
67
68
70
71
72
73
74
75
76
77
78
79
80
81
82
83

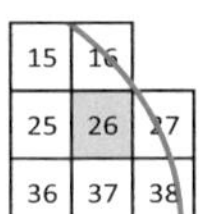

26

No.	Name	No.	Name	Name	No.	Name	No.
	Debes	45	Proclus I	Burckhardt B	1	Newcomb P	61
	[illegible]	46	[illegible]	Cleomedes		Newcomb PA	65
	Cleomedes	47	Proclus K	Cleomedes A	8	Peirce	
	Delmotte	48	Proclus M	Cleomedes B	13	Peirce C	40
	Eimmart	49	Proclus R	Cleomedes C	16	Picard	
	Rima Cleomedes	50	Yerkes E	Cleomedes D	5	Proclus	
	Rima Cleomedes I	51	Proclus S	Cleomedes DA	63	Proclus J	45
	Mare Crisium	52	Proclus T	Cleomedes DB	66	Proclus K	47
	Lacus Bonitatis	53	Proclus U	Cleomedes DE	60	Proclus L	46
	Macrobius	54	Proclus P	Cleomedes E	9	Proclus M	48
	Tisserand	55	Yerkes V	Cleomedes F	30	Proclus P	54
	Swift	56	Lyell D	Cleomedes G	26	Proclus PA	82
	Peirce	57	Proclus V	Cleomedes GA	75	Proclus R	49
	Dorsum Oppel	58	Glaisher H	Cleomedes H	31	Proclus S	51
	Fredholm	59	Glaisher F	Cleomedes J	15	Proclus T	52
	Proclus	60	Cleomedes DE	Cleomedes K	79	Proclus U	53
	Promontorium Olivium	61	Newcomb P	Cleomedes L	25	Proclus V	57
	Promontorium Lavinium	62	Hahn AD	Cleomedes M	23	Proclus W	43
	Yerkes	63	Cleomedes DA	Cleomedes N	20	Proclus X	42
	Picard	64	Eimmart TB	Cleomedes P	21	Proclus Y	44
	Curtis	65	Newcomb PA	Cleomedes Q	22	Proclus Z	41
	Crile	66	Cleomedes DB	Cleomedes R	4	Proclus ξ	81
1	Burckhardt B	67	Cleomedes α	Cleomedes S	3	Promontorium Lavinium	
2	Hahn A	68	Hahn DA	Cleomedes T	17	Promontorium Olivium	
3	Cleomedes S	69	Hahn DB	Cleomedes α	67	Rima Cleomedes	
4	Cleomedes R	70	Eimmart GB	Cleomedes β	73	Rima Cleomedes I	
5	Cleomedes D	71	Eimmart TA	Crile		Swift	
6	Debes B	72	Eimmart GA	Curtis		Tisserand	
7	Debes A	73	Cleomedes β	Debes		Tisserand A	37
8	Cleomedes A	74	Eimmart T	Debes A	7	Tisserand B	35
9	Cleomedes E	75	Cleomedes GA	Debes B	6	Tisserand D	33
10	Tralles C	76	Eimmart CA	Delmotte		Tisserand K	38
11	Tralles A	77	Eimmart FA	Dorsum Oppel		Tralles	
12	Tralles B	78	Eimmart HA	Eimmart		Tralles A	11
13	Cleomedes B	79	Cleomedes K	Eimmart C	32	Tralles B	12
14	Hahn D	80	Mare Crisium ρ	Eimmart CA	76	Tralles C	10
15	Cleomedes J	81	Proclus ξ	Eimmart F	28	Yerkes	
16	Cleomedes C	82	Proclus PA	Eimmart FA	77	Yerkes E	50
17	Cleomedes T	83	Glaisher X	Eimmart G	18	Yerkes V	55
18	Eimmart G			Eimmart GA	72		
19	Macrobius W			Eimmart GB	70		
20	Cleomedes N			Eimmart HA	78		
21	Cleomedes P			Eimmart T	74		
22	Cleomedes Q			Eimmart TA	71		
23	Cleomedes M			Eimmart TB	64		
24	Macrobius T			Fredholm			
25	Cleomedes L			Glaisher F	59		
26	Cleomedes G			Glaisher H	58		
27	Macrobius S			Glaisher X	83		
28	Eimmart F			Hahn A	2		
29	Macrobius F			Hahn AD	62		
30	Cleomedes F			Hahn D	14		
31	Cleomedes H			Hahn DA	68		
32	Eimmart C			Hahn DB	69		
33	Tisserand D			Lacus Bonitatis			
34	Macrobius C			Lyell D	56		
35	Tisserand B			Macrobius			
36	Macrobius Q			Macrobius C	34		
37	Tisserand A			Macrobius E	39		
38	Tisserand K			Macrobius F	29		
39	Macrobius E			Macrobius Q	36		
40	Peirce C			Macrobius S	27		
41	Proclus Z			Macrobius T	24		
42	Proclus X			Macrobius W	19		
43	Proclus W			Mare Crisium			
44	Proclus Y			Mare Crisium ρ	80		

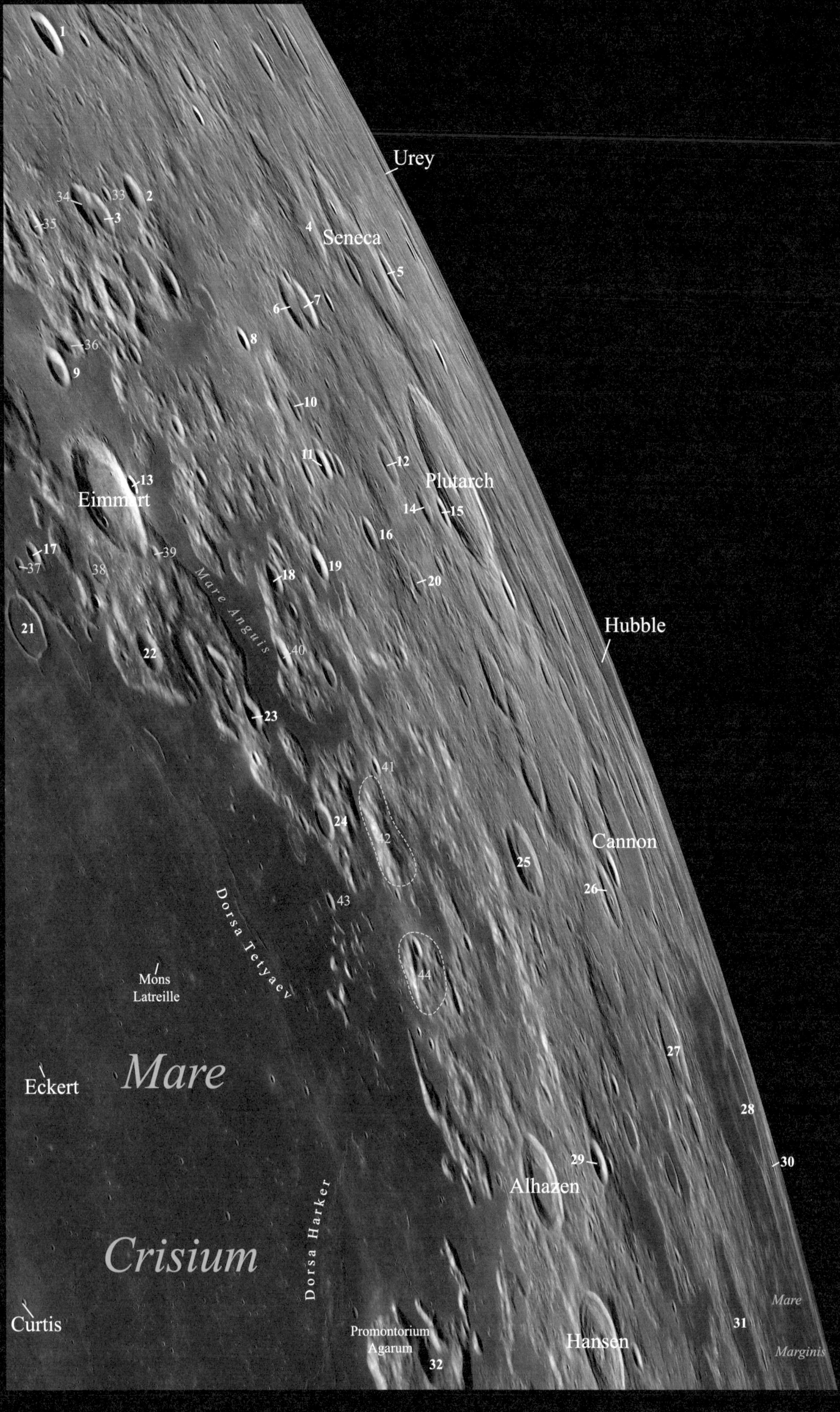

1
Urey
34 33 2
35 3
4 Seneca
5
6 7
8
36
9
10
11 12
13 Plutarch
Eimmart
14 15
16
17 39
37 38
18 19
20
Mare Anguis
Hubble
21
22 40
23
41
24
42 Cannon
25
26
43
Dorsa Tetyaev
Mons Latreille
44
27
Mare
Eckert
28
29 30
Alhazen
Crisium
Dorsa Harker
Mare
Curtis 31
Promontorium Agarum
Hansen
Marginis
32

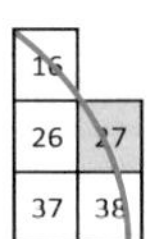

27

No.	Name
	Seneca
	Eimmart
	Plutarch
	Mare Anguis
	Hubble
	Cannon
	Dorsa Tetyaev
	Eckert
	Mons Latreille
	Alhazen
	Curtis
	Dorsa Harker
	Promontorium Agarum
	Hansen
	Mare Marginis
1	Hahn A
2	Hahn E
3	Hahn D
4	Seneca B
5	Seneca D
6	Seneca C
7	Seneca A
8	Plutarch L
9	Eimmart G
10	Plutarch K
11	Plutarch H
12	Plutarch D
13	Eimmart A
14	Plutarch N
15	Plutarch M
16	Plutarch F
17	Eimmart F
18	Eimmart D
19	Plutarch C
20	Plutarch G
21	Eimmart C
22	Eimmart H
23	Eimmart B
24	Eimmart K
25	Alhazen D
26	Cannon E
27	Cannon B
28	Goddard C
29	Alhazen A
30	Goddard B
31	Hansen B
32	Condorcet W
33	Hahn DB
34	Hahn DA
35	Eimmart TA
36	Eimmart T
37	Eimmart FA
38	Eimmart HA
39	Eimmart AB
40	Eimmart DA
41	Plutarch CA
42	Alhazen α
43	Eimmart KA
44	Alhazen β

Name	No.
Alhazen A	29
Alhazen D	25
Alhazen α	42
Alhazen β	44
Cannon	
Cannon B	27
Cannon E	26
Condorcet W	32
Curtis	
Dorsa Harker	
Dorsa Tetyaev	
Eckert	
Eimmart	
Eimmart A	13
Eimmart AB	39
Eimmart B	23
Eimmart C	21
Eimmart D	18
Eimmart DA	40
Eimmart F	17
Eimmart FA	37
Eimmart G	9
Eimmart H	22
Eimmart HA	38
Eimmart K	24
Eimmart KA	43
Eimmart T	36
Eimmart TA	35
Goddard B	30
Goddard C	28
Hahn A	1
Hahn D	3
Hahn DA	34
Hahn DB	33
Hahn E	2
Hansen	
Hansen B	31
Hubble	
Mare Anguis	
Mare Marginis	
Mons Latreille	
Plutarch	
Plutarch C	19
Plutarch CA	41
Plutarch D	12
Plutarch F	16
Plutarch G	20
Plutarch H	11
Plutarch K	10
Plutarch L	8
Plutarch M	15
Plutarch N	14
Promontorium Agarum	
Seneca	
Seneca A	7
Seneca B	4
Seneca C	6
Seneca D	5
Urey	

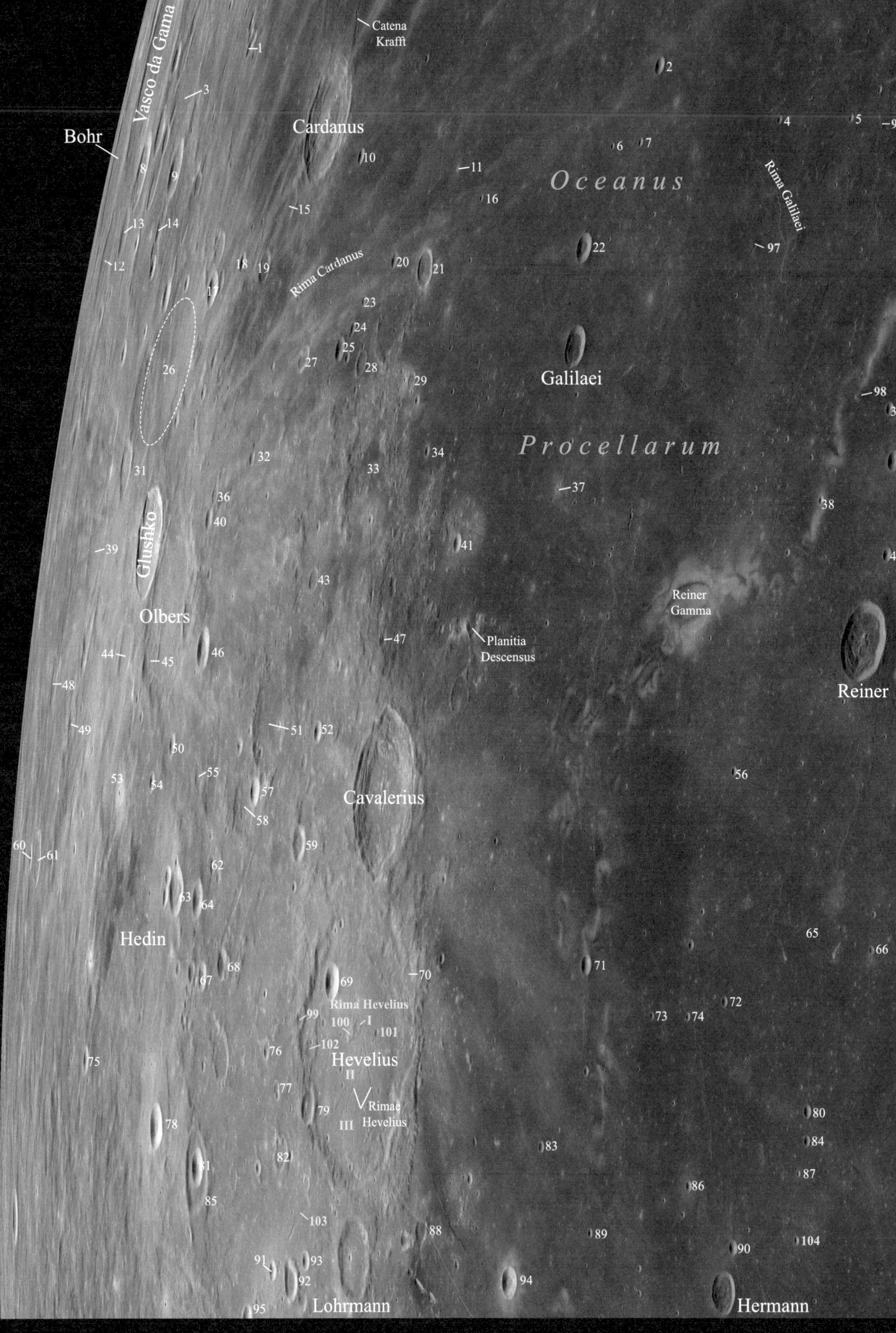

Catena Krafft
1
2
3
Vasco da Gama
Bohr
Cardanus
4
5
96
6
7
10
11
Oceanus
8
9
Rima Galilaei
16
15
13
14
22
97
12
18
19
Rima Cardanus
20
21
17
23
24
25
27
28
26
29
Galilaei
98
30
Procellarum
32
34
31
33
35
37
36
38
40
41
39
Glushko
42
43
Reiner Gamma
Olbers
47
Planitia Descensus
44
45
46
Reiner
48
49
51
52
50
55
56
53
54
57
58
Cavalerius
60
61
59
62
63
64
Hedin
65
66
68
71
67
69
70
Rima Hevelius
72
99
73
74
100
I
101
102
75
76
Hevelius
II
77
Rimae Hevelius
79
III
80
78
83
84
82
81
87
86
85
103
88
89
90
104
91
93
92
94
95
Lohrmann
Hermann

17	18
28	29
39	40

28

No.	Name	No.	Name	Name	No.	Name	No.
	Vasco da Gama	42	Reiner L	Bohr		Hermann R	87
	Catena Krafft	43	Cavalerius E	Cardanus		Hermann S	84
	Cardanus	44	Olbers K	Cardanus B	19	Hevelius	
	Bohr	45	Olbers S	Cardanus C	17	Hevelius A	69
	Rima Galilaei	46	Olbers B	Cardanus E	10	Hevelius B	79
	Oceanus Procellarum	47	Cavalerius W	Cardanus G	18	Hevelius D	71
	Rima Cardanus	48	Olbers Y	Cardanus K	1	Hevelius E	70
	Galilaei	49	Olbers W	Cardanus R	15	Hevelius F	102
	Glushko	50	Hedin S	Catena Krafft		Hevelius G	101
	Olbers	51	Cavalerius B	Cavalerius		Hevelius H	99
	Planitia Descensus	52	Cavalerius C	Cavalerius A	59	Hevelius J	82
	Reiner Gamma	53	Hedin A	Cavalerius B	51	Hevelius K	77
	Reiner	54	Hedin R	Cavalerius C	52	Hevelius L	76
	Cavalerius	55	Hedin V	Cavalerius D	33	Hevelius α	100
	Hedin	56	Reiner N	Cavalerius E	43	Lohrmann	
	Hevelius	57	Hedin L	Cavalerius F	41	Lohrmann A	94
	Rimae Hevelius	58	Hedin N	Cavalerius K	28	Lohrmann B	92
	Rima Hevelius I	59	Cavalerius A	Cavalerius L	25	Lohrmann D	88
	Rima Hevelius II	60	Hedin C	Cavalerius M	27	Lohrmann M	93
	Rima Hevelius III	61	Hedin B	Cavalerius U	29	Lohrmann N	91
	Lohrmann	62	Hedin T	Cavalerius W	47	Marius X	30
	Hermann	63	Hedin F	Cavalerius X	34	Marius ι	96
1	Cardanus K	64	Hedin G	Cavalerius Y	24	Marius σ	98
2	Galilaei E	65	Reiner R	Cavalerius Z	23	Oceanus Procellarum	
3	Vasco da Gama F	66	Reiner G	Galilaei		Olbers	
4	Galilaei L	67	Hedin K	Galilaei A	22	Olbers B	46
5	Galilaei M	68	Hedin H	Galilaei B	21	Olbers D	26
6	Galilaei K	69	Hevelius A	Galilaei D	37	Olbers G	40
7	Galilaei J	70	Hevelius E	Galilaei E	2	Olbers H	36
8	Vasco da Gama S	71	Hevelius D	Galilaei F	16	Olbers K	44
9	Vasco da Gama A	72	Hermann J	Galilaei G	11	Olbers M	39
10	Cardanus E	73	Hermann L	Galilaei H	20	Olbers N	31
11	Galilaei G	74	Hermann K	Galilaei J	7	Olbers S	45
12	Vasco da Gama C	75	Hedin Z	Galilaei K	6	Olbers V	32
13	Vasco da Gama T	76	Hevelius L	Galilaei L	4	Olbers W	49
14	Vasco da Gama P	77	Hevelius K	Galilaei M	5	Olbers Y	48
15	Cardanus R	78	Riccioli H	Galilaei χ	97	Planitia Descensus	
16	Galilaei F	79	Hevelius B	Glushko		Reiner	
17	Cardanus C	80	Hermann F	Hedin		Reiner G	66
18	Cardanus G	81	Riccioli CA	Hedin A	53	Reiner Gamma	
19	Cardanus B	82	Hevelius J	Hedin B	61	Reiner H	35
20	Galilaei H	83	Hermann H	Hedin C	60	Reiner L	42
21	Galilaei B	84	Hermann S	Hedin F	63	Reiner M	38
22	Galilaei A	85	Riccioli C	Hedin G	64	Reiner N	56
23	Cavalerius Z	86	Hermann A	Hedin H	68	Reiner R	65
24	Cavalerius Y	87	Hermann R	Hedin K	67	Riccioli C	85
25	Cavalerius L	88	Lohrmann D	Hedin L	57	Riccioli CA	81
26	Olbers D	89	Hermann C	Hedin N	58	Riccioli G	95
27	Cavalerius M	90	Hermann B	Hedin R	54	Riccioli H	78
28	Cavalerius K	91	Lohrmann N	Hedin S	50	Rima Cardanus	
29	Cavalerius U	92	Lohrmann B	Hedin T	62	Rima Galilaei	
30	Marius X	93	Lohrmann M	Hedin V	55	Rima Hevelius I	
31	Olbers N	94	Lohrmann A	Hedin Z	75	Rima Hevelius II	
32	Olbers V	95	Riccioli G	Hermann		Rima Hevelius III	
33	Cavalerius D	96	Marius ι	Hermann A	86	Rima Riccioli I	103
34	Cavalerius X	97	Galilaei χ	Hermann B	90	Rimae Hevelius	
35	Reiner H	98	Marius σ	Hermann BA	104	Vasco da Gama	
36	Olbers H	99	Hevelius H	Hermann C	89	Vasco da Gama A	9
37	Galilaei D	100	Hevelius α	Hermann F	80	Vasco da Gama C	12
38	Reiner M	101	Hevelius G	Hermann H	83	Vasco da Gama F	3
39	Olbers M	102	Hevelius F	Hermann J	72	Vasco da Gama P	14
40	Olbers G	103	Rima Riccioli I	Hermann K	74	Vasco da Gama S	8
41	Cavalerius F	104	Hermann BA	Hermann L	73	Vasco da Gama T	13

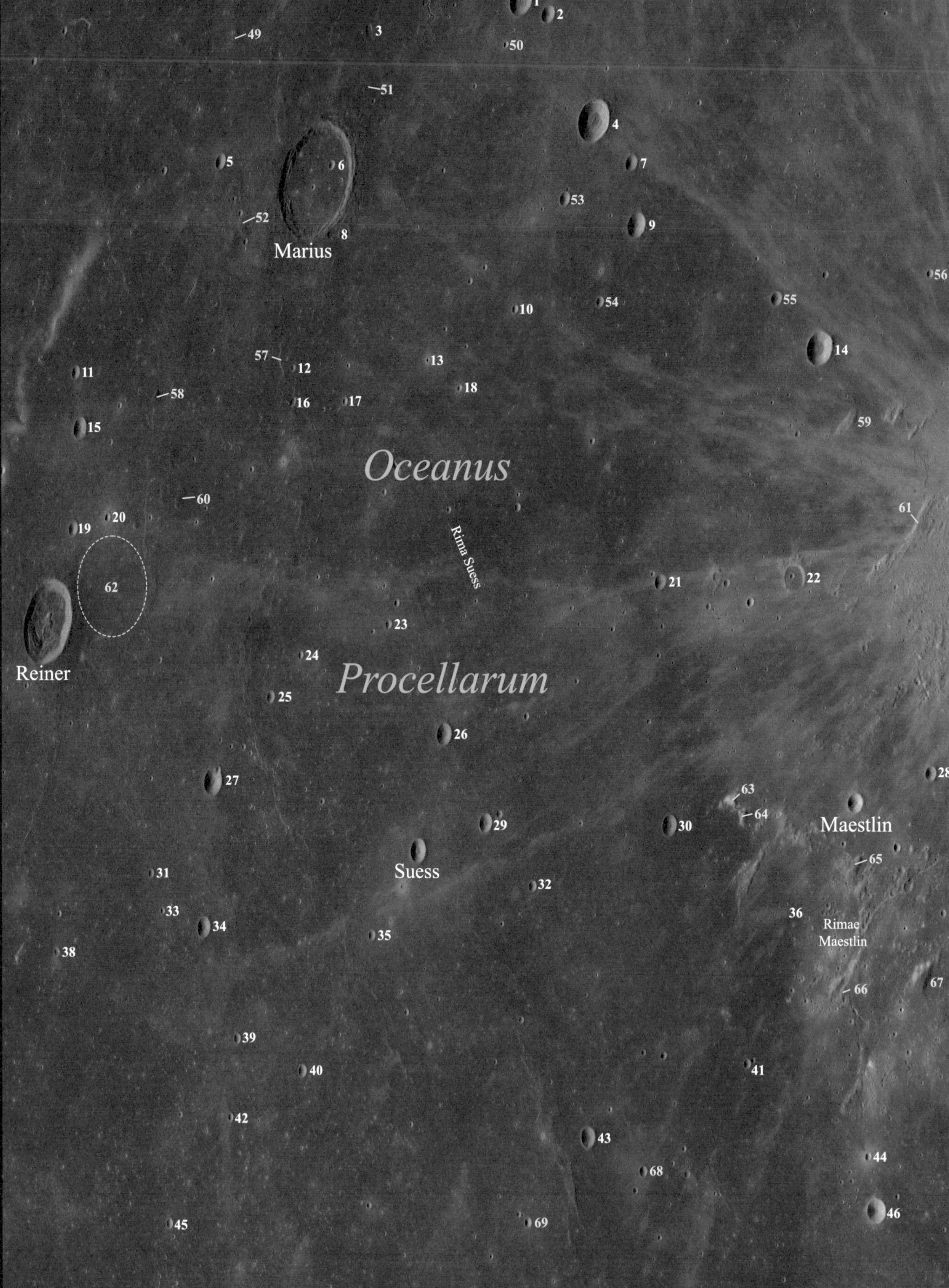

47
48
1
2
3
49
50
51
4
5
6
7
53
52
8
9
Marius
56
10
54
55
14
57
12
13
11
18
58
16
17
15
59
Oceanus
60
61
20
19
Rima Suess
62
21
22
23
Reiner
24
Procellarum
25
26
28
27
63
64
29
30
Maestlin
Suess
65
31
32
36
Rimae
Maestlin
33
34
35
37
38
66
67
39
40
41
42
43
44
68
46
45
69

29

No.	Name
	Marius
	Oceanus Procellarum
	Reiner
	Maestlin
	Suess
	Rima Suess
	Rimae Maestlin
1	Marius C
2	Marius S
3	Marius R
4	Marius A
5	Marius E
6	Marius G
7	Marius F
8	Marius H
9	Marius D
10	Marius J
11	Marius X
12	Marius Y
13	Marius V
14	Kepler C
15	Reiner H
16	Marius K
17	Marius W
18	Marius U
19	Reiner L
20	Reiner K
21	Kepler E
22	Kepler D
23	Suess J
24	Suess K
25	Suess L
26	Suess B
27	Reiner A
28	Encke J
29	Suess D
30	Maestlin H
31	Reiner U
32	Suess H
33	Reiner T
34	Reiner C
35	Suess G
36	Maestlin R
37	Encke T
38	Reiner G
39	Reiner S
40	Reiner E
41	Maestlin G
42	Reiner Q
43	Suess F
44	Encke X
45	Hermann E
46	Encke E
47	Marius γ
48	Marius μ
49	Marius ν
50	Marius CA
51	Marius κ
52	Marius ρ
53	Marius DB
54	Marius DA
55	Kepler CA
56	Kepler CB
57	Marius β
58	Marius ξ
59	Kepler π
60	Marius η
61	Kepler κ
62	Reiner P
63	Maestlin ν
64	Maestlin μ
65	Maestlin λ
66	Maestlin ι
67	Encke β
68	Suess FA
69	Suess FB

Name	No.
Encke E	46
Encke J	28
Encke T	37
Encke X	44
Encke β	67
Hermann E	45
Kepler C	14
Kepler CA	55
Kepler CB	56
Kepler D	22
Kepler E	21
Kepler κ	61
Kepler π	59
Maestlin	
Maestlin G	41
Maestlin H	30
Maestlin R	36
Maestlin ι	66
Maestlin λ	65
Maestlin μ	64
Maestlin ν	63
Marius	
Marius A	4
Marius C	1
Marius CA	50
Marius D	9
Marius DA	54
Marius DB	53
Marius E	5
Marius F	7
Marius G	6
Marius H	8
Marius J	10
Marius K	16
Marius R	3
Marius S	2
Marius U	18
Marius V	13
Marius W	17
Marius X	11
Marius Y	12
Marius β	57
Marius γ	47
Marius η	60
Marius κ	51
Marius μ	48
Marius ν	49
Marius ξ	58
Marius ρ	52
Oceanus Procellarum	
Reiner	
Reiner A	27
Reiner C	34
Reiner E	40
Reiner G	38
Reiner H	15
Reiner K	20
Reiner L	19
Reiner P	62
Reiner Q	42
Reiner S	39
Reiner T	33
Reiner U	31
Rima Suess	
Rimae Maestlin	
Suess	
Suess B	26
Suess D	29
Suess F	43
Suess FA	68
Suess FB	69
Suess G	35
Suess H	32
Suess J	23
Suess K	24
Suess L	25

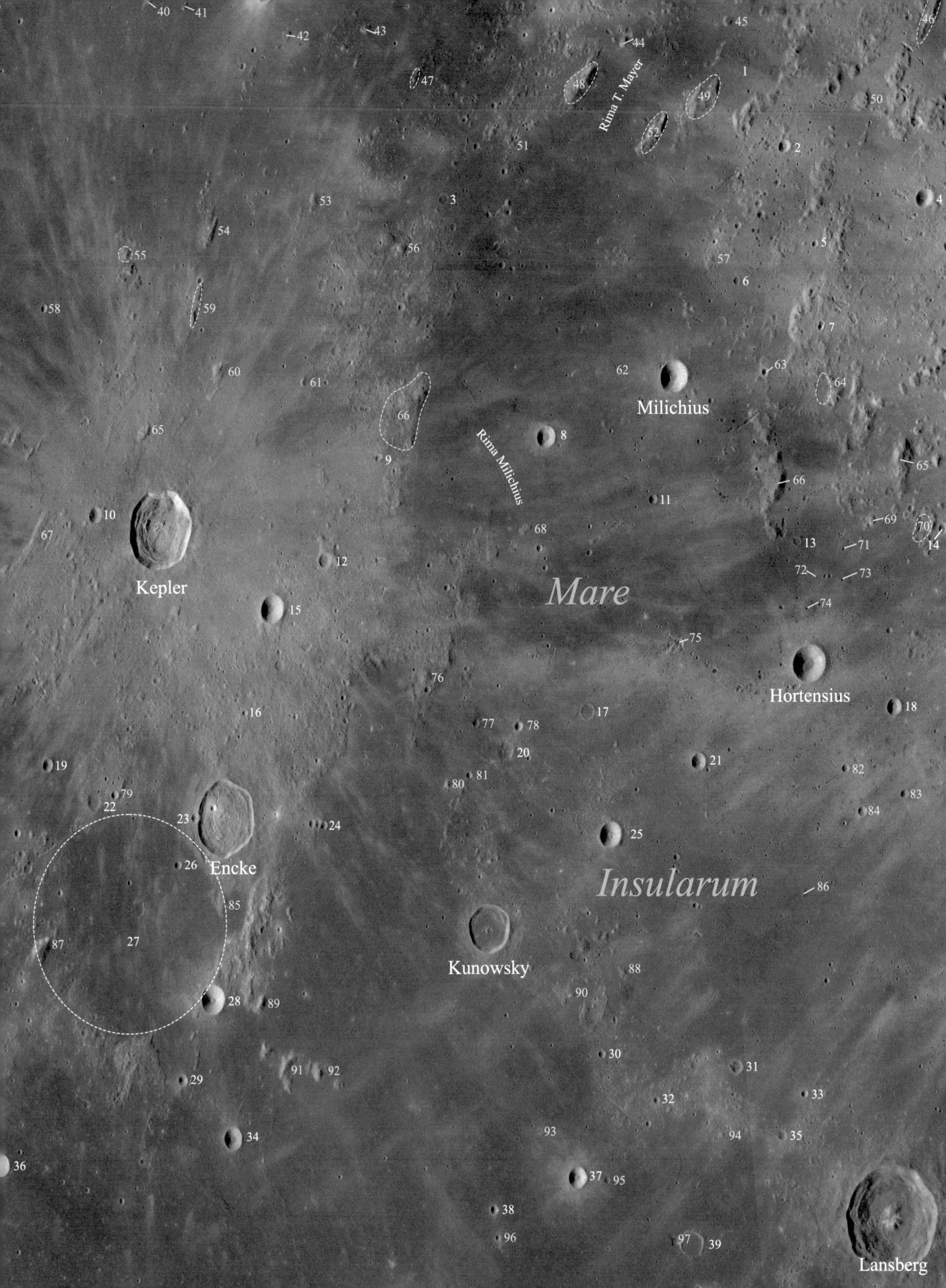
40
41
42
43
44
45
46
47
48
Rima T. Mayer
1
49
50
52
51
2
53
3
4
54
55
56
5
57
6
58
59
7
60
61
62
63
64
Milichius
66
65
Rima Milichius
8
9
65
66
10
11
67
68
69
70
13
71
14
Kepler
12
72
73
Mare
15
74
75
76
Hortensius
16
17
18
77
78
20
19
21
82
80
81
79
22
83
84
23
24
25
26
Encke
Insularum
86
85
87
27
Kunowsky
88
90
28
89
30
31
91
92
29
32
33
34
93
94
35
36
37
95
38
96
97
39
Lansberg

No.	Name	No.	Name	Name	No.	Name	No.
	Rima T. Mayer	54	Kepler ε	Bessarion ζ	41	Kunowsky CA	96
	Milichius	55	Kepler δ	Bessarion η	42	Kunowsky D	31
	Rima Milichius	56	Kepler μ	Bessarion θ	40	Kunowsky G	30
	Kepler	57	Milichius ω	Bessarion λ	47	Kunowsky H	32
	Hortensius	58	Kepler CB	Bessarion ξ	43	Kunowsky κ	88
	Mare Insularum	59	Kepler λ	Encke		Kunowsky σ	90
	Encke	60	Kepler η	Encke B	28	Kunowsky ω	80
	Kunowsky	61	Kepler φ	Encke C	34	Lansberg	
	Lansberg	62	Milichius π	Encke E	36	Lansberg A	37
1	T. Mayer P	63	Milichius BA	Encke G	22	Lansberg AA	95
2	T. Mayer F	64	Milichius B	Encke GA	79	Lansberg AB	93
3	Kepler P	65	Milichius β	Encke H	26	Lansberg G	39
4	T. Mayer D	66	Milichius α	Encke J	19	Lansberg GA	97
5	T. Mayer S	67	Kepler κ	Encke K	29	Lansberg X	33
6	Milichius C	68	Milichius τ	Encke M	24	Lansberg Y	35
7	Milichius E	69	Hortensius γ	Encke N	23	Lansberg φ	94
8	Milichius A	70	Hortensius β	Encke T	27	Mare Insularum	
9	Kepler T	71	Hortensius φ	Encke Y	16	Milichius	
10	Kepler F	72	Hortensius τ	Encke β	87	Milichius A	8
11	Milichius K	73	Hortensius σ	Encke γ	89	Milichius B	64
12	Kepler B	74	Hortensius ω	Encke θ	85	Milichius BA	63
13	Milichius D	75	Hortensius BB	Encke ρ	91	Milichius C	6
14	Hortensius G	76	Kepler θ	Encke σ	92	Milichius D	13
15	Kepler A	77	Hortensius DC	Hortensius		Milichius E	7
16	Encke Y	78	Hortensius DA	Hortensius A	25	Milichius K	11
17	Hortensius H	79	Encke GA	Hortensius B	21	Milichius α	66
18	Hortensius C	80	Kunowsky ω	Hortensius BB	75	Milichius β	65
19	Encke J	81	Hortensius DD	Hortensius C	18	Milichius κ	52
20	Hortensius D	82	Hortensius EC	Hortensius D	20	Milichius π	62
21	Hortensius B	83	Hortensius EA	Hortensius DA	78	Milichius τ	68
22	Encke G	84	Hortensius EB	Hortensius DC	77	Milichius ω	57
23	Encke N	85	Encke θ	Hortensius DD	81	Rima Milichius	
24	Encke M	86	Hortensius ρ	Hortensius EA	83	Rima T. Mayer	
25	Hortensius A	87	Encke β	Hortensius EB	84	T. Mayer D	4
26	Encke H	88	Kunowsky κ	Hortensius EC	82	T. Mayer F	2
27	Encke T	89	Encke γ	Hortensius G	14	T. Mayer P	1
28	Encke B	90	Kunowsky σ	Hortensius H	17	T. Mayer S	5
29	Encke K	91	Encke ρ	Hortensius β	70	T. Mayer α	48
30	Kunowsky G	92	Encke σ	Hortensius γ	69	T. Mayer δ	44
31	Kunowsky D	93	Lansberg AB	Hortensius ρ	86	T. Mayer ζ	49
32	Kunowsky H	94	Lansberg φ	Hortensius σ	73	T. Mayer η	46
33	Lansberg X	95	Lansberg AA	Hortensius τ	72	T. Mayer κ	51
34	Encke C	96	Kunowsky CA	Hortensius φ	71	T. Mayer ν	50
35	Lansberg Y	97	Lansberg GA	Hortensius ω	74	T. Mayer π	45
36	Encke E			Kepler			
37	Lansberg A			Kepler A	15		
38	Kunowsky C			Kepler B	12		
39	Lansberg G			Kepler CB	58		
40	Bessarion θ			Kepler F	10		
41	Bessarion ζ			Kepler P	3		
42	Bessarion η			Kepler T	9		
43	Bessarion ξ			Kepler δ	55		
44	T. Mayer δ			Kepler ε	54		
45	T. Mayer π			Kepler η	60		
46	T. Mayer η			Kepler θ	76		
47	Bessarion λ			Kepler κ	67		
48	T. Mayer α			Kepler λ	59		
49	T. Mayer ζ			Kepler μ	56		
50	T. Mayer ν			Kepler ν	53		
51	T. Mayer κ			Kepler φ	61		
52	Milichius κ			Kunowsky			
53	Kepler ν			Kunowsky C	38		

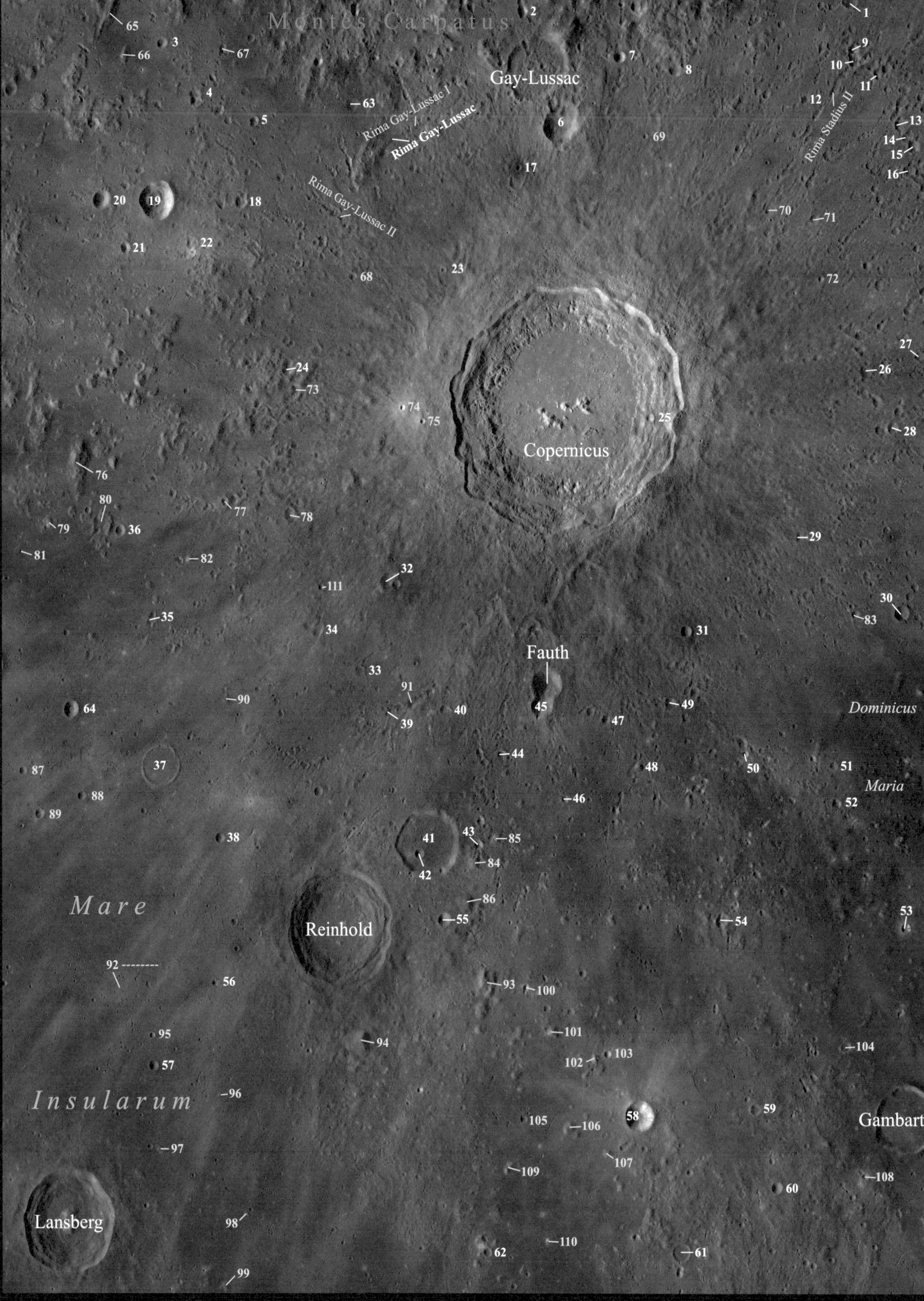
Montes Carpatus
Gay-Lussac
Rima Gay-Lussac I
Rima Gay-Lussac
Rima Gay-Lussac II
Rima Stadius II
Copernicus
Fauth
Dominicus
Maria
Mare
Insularum
Reinhold
Gambart
Lansberg
65
2
1
3
66
67
7
8
9
10
11
4
63
12
5
6
13
69
14
15
17
16
20
19
18
70
71
21
22
68
23
72
27
24
26
73
74
75
25
28
76
80
77
78
79
36
29
81
82
32
111
30
35
83
34
31
33
91
49
64
90
40
45
47
39
44
87
37
48
50
51
52
88
46
89
43
85
38
41
84
42
86
55
54
53
92
56
93
100
95
101
94
104
102
103
57
96
59
105
106
58
97
107
109
108
60
98
110
62
61
99

No.	Name	No.	Name	Name	No.	Name	No.
	Montes Carpatus	50	Fauth F	Copernicus		Gay-Lussac M	69
	Gay-Lussac	51	Fauth G	Copernicus A	25	Gay-Lussac N	17
	Rima Gay-Lussac	52	Fauth H	Copernicus B	32	Hortensius C	64
	Rima Gay-Lussac I	53	Gambart L	Copernicus BB	111	Hortensius E	37
	Rima Stadius II	54	Gambart D	Copernicus BC	78	Hortensius EA	88
	Rima Gay-Lussac II	55	Reinhold F	Copernicus BD	77	Hortensius EB	89
	Copernicus	56	Reinhold D	Copernicus C	30	Hortensius EC	87
	Fauth	57	Reinhold N	Copernicus CA	83	Hortensius F	35
	Dominicus Maria	58	Gambart A	Copernicus D	18	Hortensius G	36
	Reinhold	59	Gambart E	Copernicus DA	68	Hortensius β	80
	Mare Insularum	60	Gambart F	Copernicus E	33	Hortensius γ	79
	Gambart	61	Gambart J	Copernicus F	39	Hortensius δ	90
	Lansberg	62	Gambart R	Copernicus G	40	Hortensius ϵ	82
1	Stadius M	63	Gay-Lussac H	Copernicus GA	91	Hortensius ϕ	81
2	Gay-Lussac D	64	Hortensius C	Copernicus H	31	Lansberg	
3	T. Mayer Z	65	T. Mayer η	Copernicus J	24	Lansberg α	97
4	T. Mayer N	66	T. Mayer ω	Copernicus JC	73	Lansberg β	99
5	T. Mayer L	67	T. Mayer J	Copernicus JD	74	Lansberg λ	96
6	Gay-Lussac A	68	Copernicus DA	Copernicus JE	75	Lansberg π	98
7	Gay-Lussac F	69	Gay-Lussac M	Copernicus K	70	Mare Insularum	
8	Gay-Lussac G	70	Copernicus K	Copernicus KA	71	Milichius β	76
9	Stadius W	71	Copernicus KA	Copernicus L	12	Montes Carpatus	
10	Stadius U	72	Copernicus PA	Copernicus N	34	Reinhold	
11	Stadius J	73	Copernicus JC	Copernicus P	26	Reinhold A	42
12	Copernicus L	74	Copernicus JD	Copernicus PA	72	Reinhold B	41
13	Stadius T	75	Copernicus JE	Copernicus R	29	Reinhold C	38
14	Stadius F	76	Milichius β	Dominicus Maria		Reinhold D	56
15	Stadius S	77	Copernicus BD	Fauth		Reinhold F	55
16	Stadius E	78	Copernicus BC	Fauth A	45	Reinhold G	46
17	Gay-Lussac N	79	Hortensius γ	Fauth B	47	Reinhold H	43
18	Copernicus D	80	Hortensius β	Fauth C	48	Reinhold N	57
19	T. Mayer C	81	Hortensius ϕ	Fauth D	49	Reinhold NA	95
20	T. Mayer D	82	Hortensius ϵ	Fauth E	44	Reinhold β	94
21	T. Mayer R	83	Copernicus CA	Fauth F	50	Reinhold γ	93
22	T. Mayer H	84	Reinhold ι	Fauth G	51	Reinhold δ	92
23	Gay-Lussac J	85	Reinhold χ	Fauth H	52	Reinhold θ	86
24	Copernicus J	86	Reinhold θ	Gambart		Reinhold ι	84
25	Copernicus A	87	Hortensius EC	Gambart A	58	Reinhold χ	85
26	Copernicus P	88	Hortensius EA	Gambart AA	103	Rima Gay-Lussac	
27	Stadius D	89	Hortensius EB	Gambart AB	105	Rima Gay-Lussac I	
28	Stadius N	90	Hortensius δ	Gambart AC	100	Rima Gay-Lussac II	
29	Copernicus R	91	Copernicus GA	Gambart D	54	Rima Stadius II	
30	Copernicus C	92	Reinhold δ	Gambart E	59	Stadius D	27
31	Copernicus H	93	Reinhold γ	Gambart EA	104	Stadius E	16
32	Copernicus B	94	Reinhold β	Gambart F	60	Stadius F	14
33	Copernicus E	95	Reinhold NA	Gambart J	61	Stadius J	11
34	Copernicus N	96	Lansberg λ	Gambart L	53	Stadius M	1
35	Hortensius F	97	Lansberg α	Gambart NA	108	Stadius N	28
36	Hortensius G	98	Lansberg π	Gambart R	62	Stadius S	15
37	Hortensius E	99	Lansberg β	Gambart θ	106	Stadius T	13
38	Reinhold C	100	Gambart AC	Gambart λ	107	Stadius U	10
39	Copernicus F	101	Gambart ω	Gambart ν	102	Stadius W	9
40	Copernicus G	102	Gambart ν	Gambart ρ	109	T. Mayer C	19
41	Reinhold B	103	Gambart AA	Gambart τ	110	T. Mayer D	20
42	Reinhold A	104	Gambart EA	Gambart ω	101	T. Mayer H	22
43	Reinhold H	105	Gambart AB	Gay-Lussac		T. Mayer J	67
44	Fauth E	106	Gambart θ	Gay-Lussac A	6	T. Mayer L	5
45	Fauth A	107	Gambart λ	Gay-Lussac D	2	T. Mayer N	4
46	Reinhold G	108	Gambart NA	Gay-Lussac F	7	T. Mayer R	21
47	Fauth B	109	Gambart ρ	Gay-Lussac G	8	T. Mayer Z	3
48	Fauth C	110	Gambart τ	Gay-Lussac H	63	T. Mayer η	65
49	Fauth D	111	Copernicus BB	Gay-Lussac J	23	T. Mayer ω	66

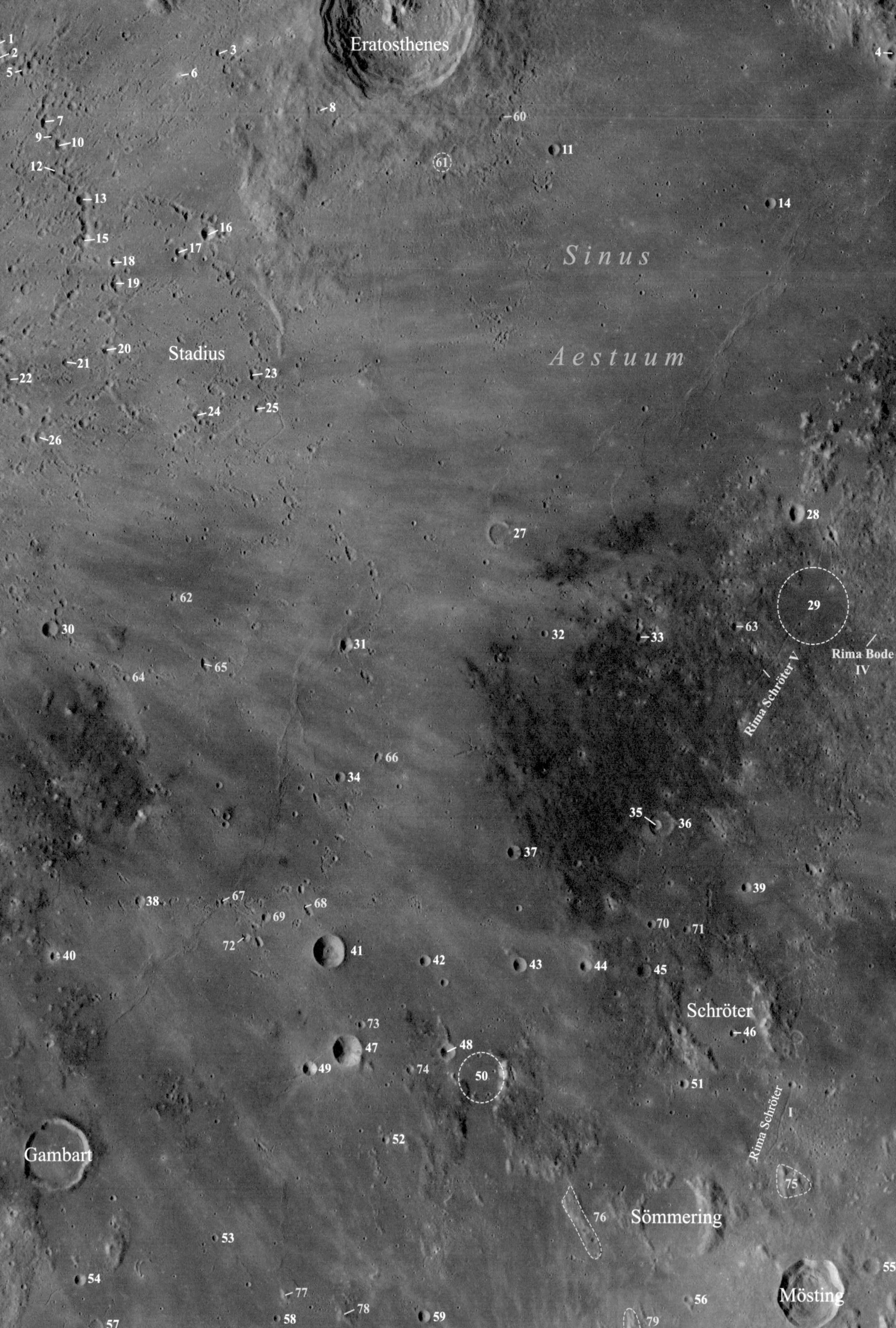
Eratosthenes
1
2
3
4
5
6
7
8
9
10
11
12
13
14
15
16
17
18
19
20
21
22
23
24
25
26
27
28
29
30
31
32
33
34
35
36
37
38
39
40
41
42
43
44
45
46
47
48
49
50
51
52
53
54
55
56
57
58
59
60
61
62
63
64
65
66
67
68
69
70
71
72
73
74
75
76
77
78
79
Sinus
Aestuum
Stadius
Rima Schröter V
Rima Bode IV
Schröter
Rima Schröter I
Gambart
Sömmering
Mösting

No.	Name	No.	Name	Name	No.	Name	No.
	[illegible]		[illegible]	Bode H	14	[illegible]	[illegible]
	Sinus Aestuum	53	Gambart S	Copernicus C	30	Sömmering β	76
	Stadius	54	Gambart N	Copernicus CB	65	Sömmering γ	75
	Rima Schröter V	55	Mösting D	Copernicus CC	62	Stadius	
	Rima Bode IV	56	Mösting K	Copernicus CD	64	Stadius A	20
	Schröter	57	Turner A	Copernicus P	22	Stadius B	16
	Gambart	58	Turner Q	Eratosthenes		Stadius C	25
	Rima Schröter	59	Turner B	Eratosthenes H	8	Stadius D	21
	Rima Schröter I	60	Eratosthenes KB	Eratosthenes K	11	Stadius E	12
	Sömmering	61	Eratosthenes KA	Eratosthenes KA	61	Stadius F	9
	Mösting	62	Copernicus CC	Eratosthenes KB	60	Stadius G	19
1	Stadius W	63	Schröter FA	Eratosthenes M	3	Stadius H	17
2	Stadius U	64	Copernicus CD	Eratosthenes Z	6	Stadius J	5
3	Eratosthenes M	65	Copernicus CB	Gambart		Stadius K	24
4	Marco Polo C	66	Gambart MA	Gambart B	47	Stadius L	23
5	Stadius J	67	Gambart CB	Gambart BB	73	Stadius N	26
6	Eratosthenes Z	68	Gambart CA	Gambart BC	74	Stadius P	15
7	Stadius T	69	Gambart CC	Gambart C	41	Stadius Q	18
8	Eratosthenes H	70	Schröter KA	Gambart CA	68	Stadius R	13
9	Stadius F	71	Schröter KB	Gambart CB	67	Stadius S	10
10	Stadius S	72	Gambart CD	Gambart CC	69	Stadius T	7
11	Eratosthenes K	73	Gambart BB	Gambart CD	72	Stadius U	2
12	Stadius E	74	Gambart BC	Gambart G	49	Stadius W	1
13	Stadius R	75	Sömmering γ	Gambart H	42	Turner A	57
14	Bode H	76	Sömmering β	Gambart K	38	Turner B	59
15	Stadius P	77	Turner μ	Gambart L	40	Turner Q	58
16	Stadius B	78	Turner λ	Gambart M	34	Turner λ	78
17	Stadius H	79	Mösting δ	Gambart MA	66	Turner μ	77
18	Stadius Q			Gambart N	54		
19	Stadius G			Gambart S	53		
20	Stadius A			Marco Polo C	4		
21	Stadius D			Mösting			
22	Copernicus P			Mösting D	55		
23	Stadius L			Mösting K	56		
24	Stadius K			Mösting δ	79		
25	Stadius C			Rima Bode IV			
26	Stadius N			Rima Schröter			
27	Schröter C			Rima Schröter I			
28	Schröter J			Rima Schröter V			
29	Schröter F			Schröter			
30	Copernicus C			Schröter A	35		
31	Schröter M			Schröter C	27		
32	Schröter S			Schröter D	37		
33	Schröter T			Schröter E	46		
34	Gambart M			Schröter F	29		
35	Schröter A			Schröter FA	63		
36	Schröter W			Schröter G	43		
37	Schröter D			Schröter H	44		
38	Gambart K			Schröter J	28		
39	Schröter U			Schröter K	45		
40	Gambart L			Schröter KA	70		
41	Gambart C			Schröter KB	71		
42	Gambart H			Schröter L	51		
43	Schröter G			Schröter M	31		
44	Schröter H			Schröter S	32		
45	Schröter K			Schröter T	33		
46	Schröter E			Schröter U	39		
47	Gambart B			Schröter W	36		
48	Sömmering P			Sinus Aestuum			
49	Gambart G			Sömmering			
50	Sömmering R			Sömmering A	52		
51	Schröter L			Sömmering P	48		

Mare
Vaporum
Rimae Bode
II
Rimae Bode
I
III
Rima Hyginus
Ukert
IV
Rimae Bode
Bode
Pallas
Murchison
Chladni
Triesnecker
Rimae Triesnecker
VI
VII
II
I
V
Rima Rhaeticus I
Sinus Medii
Bruce
Blagg
Rhaeticus
Oppolzer
1
2
3
4
5
6
7
8
9
10
11
12
13
14
15
16
17
18
19
20
21
22
23
24
25
26
27
28
29
30
31
32
33
34
35
36
37
38
39
40
41
42
43
44
45
46
47
48
49
50
51
52
53
54
55
56
57
58
59
60
61
62
63

21	22	23
32	33	34
43	44	45

33

No.	Name	No.	Name	Name	No.	Name	No.
	Mare Vaporum	37	Triesnecker F	Blagg		Rima Triesnecker VI	
	Rimae Bode	38	Pallas W	Bode		Rima Triesnecker VII	
	Rima Bode I	39	Triesnecker G	Bode A	12	Rimae Bode	
	Rima Bode II	40	Pallas F	Bode B	11	Rimae Triesnecker	
	Rima Bode III	41	Triesnecker J	Bode BA	57	Sinus Medii	
	Rima Bode IV	42	Triesnecker H	Bode C	3	Sömmering ζ	60
	Rima Hyginus	43	Pallas D	Bode D	23	Triesnecker	
	Ukert	44	Pallas V	Bode E	4	Triesnecker E	30
	Bode	45	Rhaeticus A	Bode EA	56	Triesnecker F	37
	Pallas	46	Rhaeticus M	Bode G	25	Triesnecker G	39
	Murchison	47	Rhaeticus N	Bode K	9	Triesnecker H	42
	Chladni	48	Mösting E	Bode L	29	Triesnecker J	41
	Triesnecker	49	Rhaeticus L	Bode N	6	Ukert	
	Rimae Triesnecker	50	Mösting D	Bruce		Ukert A	15
	Rima Triesnecker I	51	Oppolzer A	Chladni		Ukert B	17
	Rima Triesnecker II	52	Réaumur D	Hyginus A	27	Ukert D	58
	Rima Triesnecker III	53	Mösting L	Hyginus B	20	Ukert E	13
	Rima Triesnecker V	54	Rhaeticus J	Hyginus D	5	Ukert J	7
	Rima Triesnecker VI	55	Rhaeticus H	Marco Polo C	1	Ukert K	26
	Rima Triesnecker VII	56	Bode EA	Marco Polo T	2	Ukert M	19
	Sinus Medii	57	Bode BA	Mare Vaporum		Ukert N	22
	Bruce	58	Ukert D	Mösting D	50	Ukert P	21
	Blagg	59	Pallas FA	Mösting E	48	Ukert R	18
	Rima Rhaeticus I	60	Sömmering ζ	Mösting L	53	Ukert V	16
	Rhaeticus	61	Pallas ϕ	Murchison		Ukert W	10
	Oppolzer	62	Pallas η	Murchison T	34	Ukert X	14
1	Marco Polo C	63	Pallas ω	Oppolzer		Ukert Y	8
2	Marco Polo T			Oppolzer A	51		
3	Bode C			Pallas			
4	Bode E			Pallas A	28		
5	Hyginus D			Pallas B	35		
6	Bode N			Pallas C	33		
7	Ukert J			Pallas D	43		
8	Ukert Y			Pallas E	36		
9	Bode K			Pallas F	40		
10	Ukert W			Pallas FA	59		
11	Bode B			Pallas H	32		
12	Bode A			Pallas N	24		
13	Ukert E			Pallas V	44		
14	Ukert X			Pallas W	38		
15	Ukert A			Pallas X	31		
16	Ukert V			Pallas η	62		
17	Ukert B			Pallas ϕ	61		
18	Ukert R			Pallas ω	63		
19	Ukert M			Réaumur D	52		
20	Hyginus B			Rhaeticus			
21	Ukert P			Rhaeticus A	45		
22	Ukert N			Rhaeticus H	55		
23	Bode D			Rhaeticus J	54		
24	Pallas N			Rhaeticus L	49		
25	Bode G			Rhaeticus M	46		
26	Ukert K			Rhaeticus N	47		
27	Hyginus A			Rima Bode I			
28	Pallas A			Rima Bode II			
29	Bode L			Rima Bode III			
30	Triesnecker E			Rima Bode IV			
31	Pallas X			Rima Hyginus			
32	Pallas H			Rima Rhaeticus I			
33	Pallas C			Rima Triesnecker I			
34	Murchison T			Rima Triesnecker II			
35	Pallas B			Rima Triesnecker III			
36	Pallas E			Rima Triesnecker V			

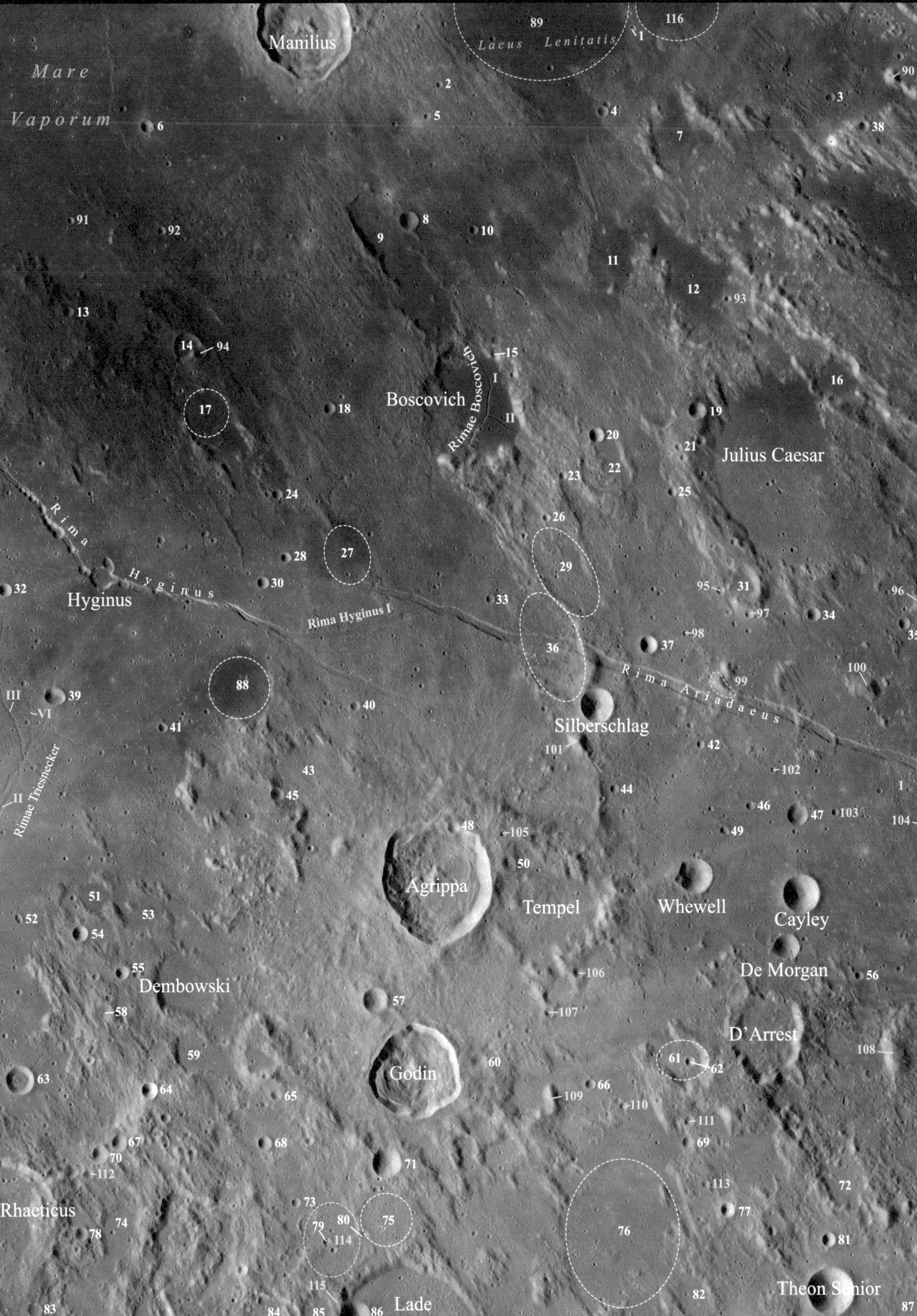
Manilius
Mare Vaporum
Lacus Lenitatis
89
116
1
90
2
3
4
5
6
7
38
8
9
10
91
92
11
12
93
13
14
94
15
Rimae Boscovich
I
II
16
17
18
Boscovich
19
20
21
Julius Caesar
22
23
24
25
26
Rima Hyginus
27
28
29
30
32
Hyginus
33
Rima Hyginus I
31
95
96
97
34
35
36
37
98
88
39
III
VI
40
41
Silberschlag
99
100
Rima Ariadaeus
101
42
102
Rimae Triesnecker
II
43
44
45
46
47
103
104
I
48
105
49
50
Agrippa
Tempel
Whewell
Cayley
51
52
53
54
De Morgan
55
Dembowski
106
56
57
58
107
D'Arrest
59
60
61
62
108
63
Godin
64
65
66
109
110
111
67
68
69
70
71
112
113
72
73
Rhaeticus
74
75
79
80
114
76
77
78
81
115
82
Theon Senior
83
84
85
86
Lade
87

No.	Name	No.	Name	Name	No.	Name	No.
	Manilius	45	Agrippa E	Agrippa		Julius Caesar Q	7
	Mare Vaporum	46	Whewell B	Agrippa B	40	Julius Caesar η	96
	Lacus Lenitatis	47	Ariadaeus B	Agrippa D	53	Lacus Lenitatis	
	Boscovich	48	Agrippa H	Agrippa E	45	Lade	
	Rimae Boscovich	49	Whewell A	Agrippa F	50	Lade A	76
	Rima Boscovich I	50	Agrippa F	Agrippa FA	105	Lade B	75
	Rima Boscovich II	51	Agrippa G	Agrippa G	51	Lade C	114
	Julius Caesar	52	Triesnecker G	Agrippa H	48	Lade D	82
	Rima Hyginus	53	Agrippa D	Agrippa S	43	Lade M	86
	Rima Hyginus I	54	Triesnecker D	Ariadaeus B	47	Lade S	84
	Hyginus	55	Dembowski A	Ariadaeus BA	103	Lade T	85
	Rima Ariadaeus	56	Dionysius B	Ariadaeus DA	104	Lade U	80
	Silberschlag	57	Godin A	Ariadaeus α	99	Lade V	79
	Rimae Triesnecker	58	Dembowski B	Ariadaeus ϵ	100	Lade W	73
	Rima Triesnecker II	59	Dembowski C	Boscovich		Lade λ	115
	Rima Triesnecker III	60	Godin G	Boscovich A	20	Manilius	
	Rima Triesnecker VI	61	D'Arrest M	Boscovich B	18	Manilius C	8
	Agrippa	62	D'Arrest A	Boscovich C	26	Manilius D	6
	Tempel	63	Rhaeticus A	Boscovich D	23	Manilius DA	92
	Whewell	64	Rhaeticus B	Boscovich E	22	Manilius DB	91
	Cayley	65	Godin C	Boscovich F	15	Manilius K	10
	De Morgan	66	Godin E	Boscovich P	9	Manilius N	89
	D'Arrest	67	Rhaeticus G	Cayley		Manilius T	5
	Dembowski	68	Godin D	D'Arrest		Manilius U	2
	Godin	69	D'Arrest B	D'Arrest A	62	Manilius W	4
	Rhaeticus	70	Rhaeticus D	D'Arrest B	69	Manilius X	1
	Lade	71	Godin B	D'Arrest BA	111	Mare Vaporum	
	Theon Senior	72	D'Arrest R	D'Arrest BC	110	Menelaus D	38
1	Manilius X	73	Lade W	D'Arrest M	61	Menelaus E	3
2	Manilius U	74	Rhaeticus F	D'Arrest R	72	Menelaus R	116
3	Menelaus E	75	Lade B	De Morgan		Menelaus α	90
4	Manilius W	76	Lade A	Delambre H	87	Rhaeticus	
5	Manilius T	77	Theon Senior B	Dembowski		Rhaeticus A	63
6	Manilius D	78	Rhaeticus E	Dembowski A	55	Rhaeticus B	64
7	Julius Caesar Q	79	Lade V	Dembowski B	58	Rhaeticus D	70
8	Manilius C	80	Lade U	Dembowski C	59	Rhaeticus DA	112
9	Boscovich P	81	Theon Senior A	Dionysius B	56	Rhaeticus E	78
10	Manilius K	82	Lade D	Dionysius γ	108	Rhaeticus F	74
11	Julius Caesar F	83	Rhaeticus H	Godin		Rhaeticus G	67
12	Julius Caesar P	84	Lade S	Godin A	57	Rhaeticus H	83
13	Hyginus G	85	Lade T	Godin B	71	Rima Ariadaeus	
14	Hyginus N	86	Lade M	Godin C	65	Rima Boscovich I	
15	Boscovich F	87	Delambre H	Godin D	68	Rima Boscovich II	
16	Julius Caesar G	88	Hyginus S	Godin E	66	Rima Hyginus	
17	Hyginus W	89	Manilius N	Godin G	60	Rima Hyginus I	
18	Boscovich B	90	Menelaus α	Godin β	109	Rima Triesnecker II	
19	Julius Caesar B	91	Manilius DB	Hyginus		Rima Triesnecker III	
20	Boscovich A	92	Manilius DA	Hyginus A	39	Rima Triesnecker VI	
21	Julius Caesar J	93	Julius Caesar PA	Hyginus B	32	Rimae Boscovich	
22	Boscovich E	94	Hyginus NA	Hyginus C	30	Rimae Triesnecker	
23	Boscovich D	95	Julius Caesar AB	Hyginus E	24	Silberschlag	
24	Hyginus E	96	Julius Caesar η	Hyginus F	28	Silberschlag A	37
25	Julius Caesar H	97	Julius Caesar AC	Hyginus G	13	Silberschlag AB	98
26	Boscovich C	98	Silberschlag AB	Hyginus H	41	Silberschlag D	33
27	Hyginus Z	99	Ariadaeus α	Hyginus N	14	Silberschlag E	44
28	Hyginus F	100	Ariadaeus ϵ	Hyginus NA	94	Silberschlag G	42
29	Silberschlag S	101	Silberschlag β	Hyginus S	88	Silberschlag P	36
30	Hyginus C	102	Whewell BA	Hyginus W	17	Silberschlag S	29
31	Julius Caesar A	103	Ariadaeus BA	Hyginus Z	27	Silberschlag β	101
32	Hyginus B	104	Ariadaeus DA	Julius Caesar		Tempel	
33	Silberschlag D	105	Agrippa FA	Julius Caesar A	31	Tempel A	106
34	Julius Caesar C	106	Tempel A	Julius Caesar AB	95	Tempel AB	107
35	Julius Caesar D	107	Tempel AB	Julius Caesar AC	97	Theon Senior	
36	Silberschlag P	108	Dionysius γ	Julius Caesar B	19	Theon Senior A	81
37	Silberschlag A	109	Godin β	Julius Caesar C	34	Theon Senior B	77
38	Menelaus D	110	D'Arrest BC	Julius Caesar D	35	Theon Senior BA	113
39	Hyginus A	111	D'Arrest BA	Julius Caesar F	11	Triesnecker D	54
40	Agrippa B	112	Rhaeticus DA	Julius Caesar G	16	Triesnecker G	52
41	Hyginus H	113	Theon Senior BA	Julius Caesar H	25	Whewell	
42	Silberschlag G	114	Lade C	Julius Caesar J	21	Whewell A	49
43	Agrippa S	115	Lade λ	Julius Caesar P	12	Whewell B	46
44	Silberschlag E	116	Menelaus R	Julius Caesar PA	93	Whewell BA	102

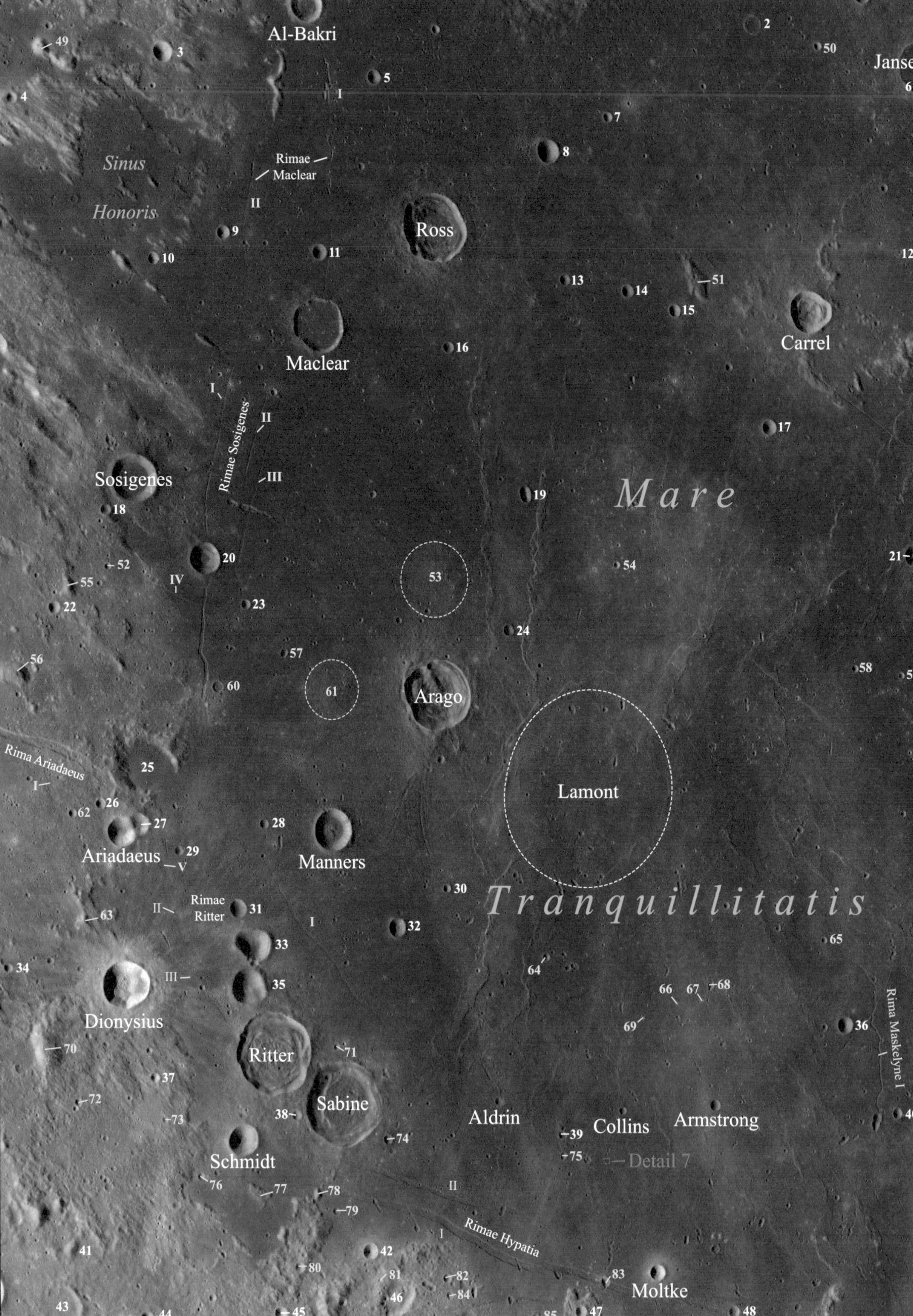

Al-Bakri
Jansen
Sinus
Honoris
Rimae
Maclear
Ross
Carrel
Maclear
Rimae Sosigenes
Sosigenes
Mare
Arago
Lamont
Rima Ariadaeus
Ariadaeus
Manners
Rimae
Ritter
Tranquillitatis
Dionysius
Ritter
Sabine
Aldrin
Collins
Armstrong
Rima Maskelyne I
Schmidt
Detail 7
Rimae Hypatia
Moltke

No.	Name
	Al-Bakri
	Jansen
	Rimae Maclear
	Sinus Honoris
	Ross
	Maclear
	Carrel
	Sosigenes
	Rimae Sosigenes
	Rima Sosigenes I
	Rima Sosigenes II
	Rima Sosigenes III
	Rima Sosigenes IV
	Arago
	Lamont
	Mare Tranquillitatis
	Rima Ariadaeus
	Rima Ariadaeus I
	Ariadaeus
	Manners
	Rimae Ritter
	Rima Ritter I
	Rima Ritter II
	Rima Ritter III
	Rima Ritter V
	Dionysius
	Ritter
	Sabine
	Rima Maskelyne I
	Aldrin
	Collins
	Armstrong
	Schmidt
	Rimae Hypatia
	Rima Hypatia I
	Rima Hypatia II
	Moltke
1	Jansen E
2	Plinius B
3	Auwers A
4	Menelaus D
5	Tacquet C
6	Jansen Y
7	Plinius A
8	Ross D
9	Ross C
10	Maclear A
11	Ross B
12	Jansen H
13	Ross E
14	Ross F
15	Ross G
16	Ross H
17	Jansen G
18	Sosigenes B
19	Arago E
20	Sosigenes A
21	Maskelyne M
22	Julius Caesar D
23	Sosigenes C
24	Arago D
25	Ariadaeus E
26	Ariadaeus D
27	Ariadaeus A
28	Manners A
29	Ariadaeus F
30	Arago C
31	Ritter D
32	Arago B
33	Ritter B
34	Dionysius B
35	Ritter C
36	Maskelyne G
37	Dionysius A
38	Sabine A
39	Sabine C
40	Maskelyne X
41	Delambre J
42	Hypatia E
43	Delambre H
44	Delambre D
45	Delambre F
46	Hypatia C
47	Moltke A
48	Moltke B
49	Menelaus α
50	Jansen EA
51	Ross μ
52	Sosigenes BA
53	Arago α
54	Arago EA
55	Julius Caesar η
56	Ariadaeus ϵ
57	Sosigenes CB
58	Maskelyne MA
59	Maskelyne MB
60	Sosigenes CA
61	Arago β
62	Ariadaeus DA
63	Dionysius β
64	Arago CA
65	Maskelyne GA
66	Sabine EF
67	Sabine EB
68	Sabine EA
69	Sabine DM
70	Dionysius γ
71	Sabine AC
72	Dionysius AB
73	Dionysius AC
74	Sabine AD
75	Sabine CA
76	Schmidt A
77	Sabine β
78	Sabine α
79	Sabine AB
80	Delambre FA
81	Hypatia CC
82	Hypatia CA
83	Moltke AC
84	Hypatia CD
85	Moltke AD

Name	No.
Al-Bakri	
Aldrin	
Arago	
Arago B	32
Arago C	30
Arago CA	64
Arago D	24
Arago E	19
Arago EA	54
Arago α	53
Arago β	61
Ariadaeus	
Ariadaeus A	27
Ariadaeus D	26
Ariadaeus DA	62
Ariadaeus E	25
Ariadaeus F	29
Ariadaeus ϵ	56
Armstrong	
Auwers A	3
Carrel	
Collins	
Delambre D	44
Delambre F	45
Delambre FA	80
Delambre H	43
Delambre J	41
Dionysius	
Dionysius A	37
Dionysius AB	72
Dionysius AC	73
Dionysius B	34
Dionysius β	63
Dionysius γ	70
Hypatia C	46
Hypatia CA	82
Hypatia CC	81
Hypatia CD	84
Hypatia E	42
Jansen	
Jansen E	1
Jansen EA	50
Jansen G	17
Jansen H	12
Jansen Y	6
Julius Caesar D	22
Julius Caesar η	55
Lamont	
Maclear	
Maclear A	10
Manners	
Manners A	28
Mare Tranquillitatis	
Maskelyne G	36
Maskelyne GA	65
Maskelyne M	21
Maskelyne MA	58
Maskelyne MB	59
Maskelyne X	40
Menelaus D	4
Menelaus α	49
Moltke	
Moltke A	47
Moltke AC	83
Moltke AD	85
Moltke B	48
Plinius A	7
Plinius B	2
Rima Ariadaeus	
Rima Ariadaeus I	
Rima Hypatia I	
Rima Hypatia II	
Rima Maskelyne I	
Rima Ritter I	
Rima Ritter II	
Rima Ritter III	
Rima Ritter V	
Rima Sosigenes I	
Rima Sosigenes II	
Rima Sosigenes III	
Rima Sosigenes IV	
Rimae Hypatia	
Rimae Maclear	
Rimae Ritter	
Rimae Sosigenes	
Ritter	
Ritter B	33
Ritter C	35
Ritter D	31
Ross	
Ross B	11
Ross C	9
Ross D	8
Ross E	13
Ross F	14
Ross G	15
Ross H	16
Ross μ	51
Sabine	
Sabine A	38
Sabine AB	79
Sabine AC	71
Sabine AD	74
Sabine C	39
Sabine CA	75
Sabine DM	69
Sabine EA	68
Sabine EB	67
Sabine EF	66
Sabine α	78
Sabine β	77
Schmidt	
Schmidt A	76
Sinus Honoris	
Sosigenes	
Sosigenes A	20
Sosigenes B	18
Sosigenes BA	52
Sosigenes C	23
Sosigenes CA	60
Sosigenes CB	57
Tacquet C	5

Mons Esam
Rima Jansen
Diana
Grace
Lucian
Jansen
Lyell
Cajal
Rima Cauchy I
Mare
Cauchy
Rupes Cauchy
Sinas
Donna
Tranquillitatis
Aryabhata
Zähringer
Wallach
Menzel
Rima Maskelyne I
Maskelyne
Mount Marilyn
Censorinus

24	25	26
35	36	37
46	47	48

36

No.	Name	No.	Name	Name	No.	Name	No.
	Mons Esam	39	Maskelyne A	Aryabhata		Maskelyne PA	74
	Rima Jansen	40	Maskelyne T	Cajal		Maskelyne PB	73
	Rima Jansen I	41	Lubbock M	Cauchy		Maskelyne R	29
	Diana	42	Censorinus V	Cauchy A	8	Maskelyne T	40
	Grace	43	Censorinus X	Cauchy B	18	Maskelyne TA	81
	Lucian	44	Censorinus K	Cauchy C	21	Maskelyne W	36
	Jansen	45	Censorinus J	Cauchy D	15	Maskelyne X	34
	Lyell	46	Censorinus W	Cauchy E	20	Maskelyne Y	33
	Cajal	47	Jansen δ	Cauchy F	19	Maskelyne α	77
	Rima Cauchy	48	Sinas β	Cauchy M	24	Maskelyne ε	60
	Rima Cauchy I	49	Cauchy V	Cauchy V	49	Maskelyne ζ	75
	Cauchy	50	Cauchy τ	Cauchy ρ	52	Maskelyne η	68
	Rupes Cauchy	51	Sinas α	Cauchy τ	50	Maskelyne θ	76
	Sinas	52	Cauchy ρ	Censorinus		Maskelyne ι	56
	Mare Tranquillitatis	53	Maskelyne MA	Censorinus A	38	Maskelyne κ	59
	Donna	54	Maskelyne MB	Censorinus AB	83	Maskelyne λ	63
	Aryabhata	55	Maskelyne NA	Censorinus CA	80	Maskelyne ϕ	79
	Zähringer	56	Maskelyne ι	Censorinus J	45	Menzel	
	Wallach	57	Taruntius EC	Censorinus K	44	Mons Esam	
	Menzel	58	Taruntius ED	Censorinus V	42	Mount Marilyn	
	Rima Maskelyne I	59	Maskelyne κ	Censorinus W	46	Rima Cauchy	
	Maskelyne	60	Maskelyne ε	Censorinus X	43	Rima Cauchy I	
	Mount Marilyn	61	Maskelyne HA	Diana		Rima Jansen	
	Censorinus	62	Maskelyne FD	Donna		Rima Jansen I	
1	Jansen E	63	Maskelyne λ	Grace		Rima Maskelyne I	
2	Lyell B	64	Taruntius θ	Jansen		Rupes Cauchy	
3	Lyell A	65	Taruntius κ	Jansen E	1	Secchi α	82
4	Vitruvius G	66	Maskelyne KC	Jansen H	9	Secchi β	78
5	Jansen Y	67	Maskelyne KB	Jansen K	10	Sinas	
6	Maraldi W	68	Maskelyne η	Jansen T	11	Sinas A	23
7	Jansen U	69	Taruntius ι	Jansen U	7	Sinas E	16
8	Cauchy A	70	Maskelyne JA	Jansen W	12	Sinas G	17
9	Jansen H	71	Maskelyne KA	Jansen Y	5	Sinas H	14
10	Jansen K	72	Maskelyne DB	Jansen δ	47	Sinas J	13
11	Jansen T	73	Maskelyne PB	Lubbock M	41	Sinas K	25
12	Jansen W	74	Maskelyne PA	Lucian		Sinas α	51
13	Sinas J	75	Maskelyne ζ	Lyell		Sinas β	48
14	Sinas H	76	Maskelyne θ	Lyell A	3	Taruntius EC	57
15	Cauchy D	77	Maskelyne α	Lyell B	2	Taruntius ED	58
16	Sinas E	78	Secchi β	Maraldi W	6	Taruntius θ	64
17	Sinas G	79	Maskelyne ϕ	Mare Tranquillitatis		Taruntius ι	69
18	Cauchy B	80	Censorinus CA	Maskelyne		Taruntius κ	65
19	Cauchy F	81	Maskelyne TA	Maskelyne A	39	Vitruvius G	4
20	Cauchy E	82	Secchi α	Maskelyne B	32	Wallach	
21	Cauchy C	83	Censorinus AB	Maskelyne C	35	Zähringer	
22	Maskelyne M			Maskelyne D	31		
23	Sinas A			Maskelyne DB	72		
24	Cauchy M			Maskelyne F	27		
25	Sinas K			Maskelyne FD	62		
26	Maskelyne N			Maskelyne HA	61		
27	Maskelyne F			Maskelyne J	30		
28	Maskelyne K			Maskelyne JA	70		
29	Maskelyne R			Maskelyne K	28		
30	Maskelyne J			Maskelyne KA	71		
31	Maskelyne D			Maskelyne KB	67		
32	Maskelyne B			Maskelyne KC	66		
33	Maskelyne Y			Maskelyne M	22		
34	Maskelyne X			Maskelyne MA	53		
35	Maskelyne C			Maskelyne MB	54		
36	Maskelyne W			Maskelyne N	26		
37	Maskelyne P			Maskelyne NA	55		
38	Censorinus A			Maskelyne P	37		

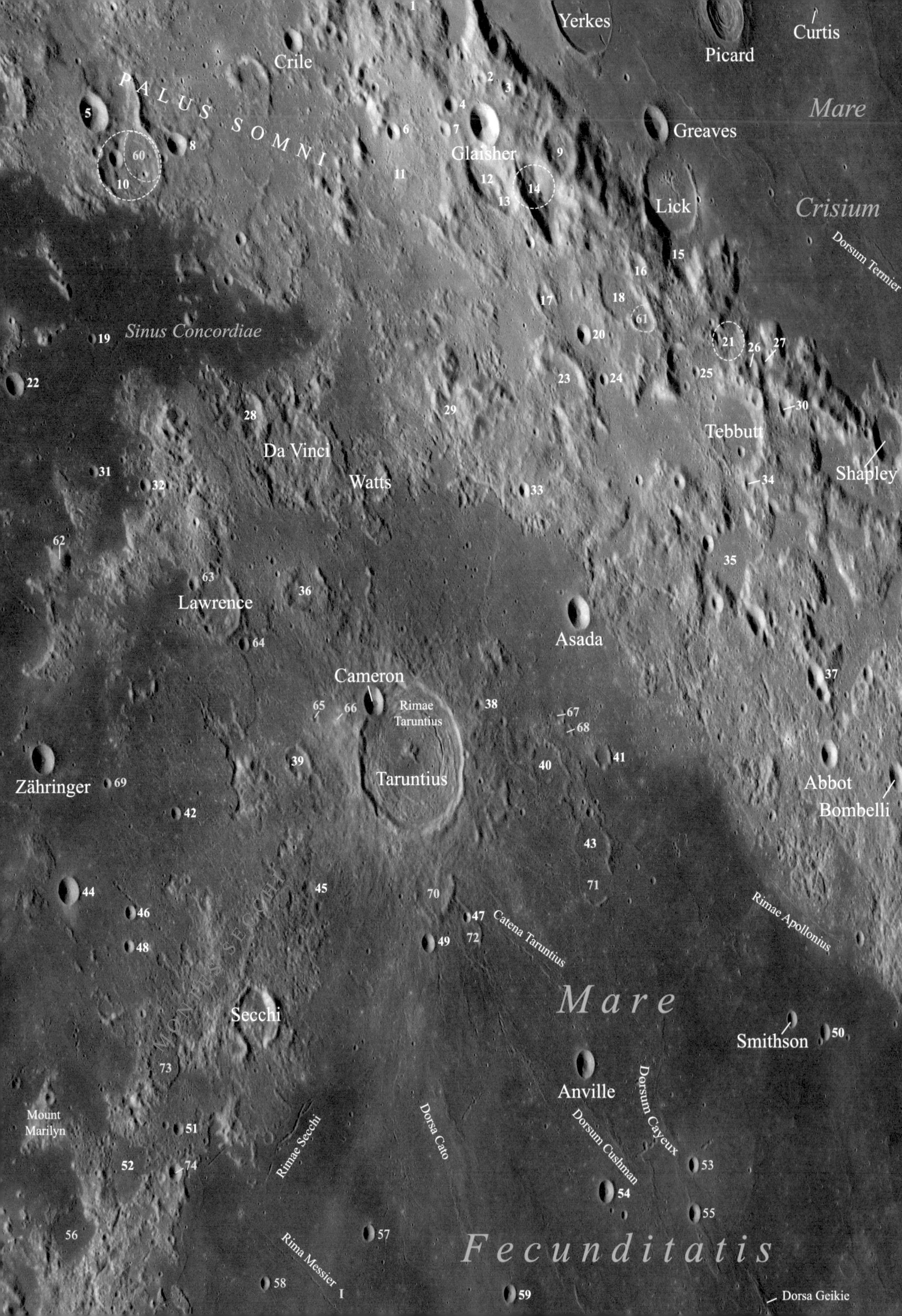

Yerkes
Picard
Curtis
Crile
Mare
Crisium
PALUS SOMNI
Greaves
Glaisher
Lick
Dorsum Termier
Sinus Concordiae
Tebbutt
Shapley
Da Vinci
Watts
Lawrence
Asada
Cameron
Rimae Taruntius
Taruntius
Zähringer
Abbot
Bombelli
Catena Taruntius
Rimae Apollonius
MONTES SECCHI
Secchi
Mare
Smithson
Anville
Dorsum Cayeux
Dorsum Cushman
Dorsa Cato
Rimae Secchi
Mount Marilyn
Rima Messier
Fecunditatis
Dorsa Geikie

No.	Name	No.	Name	Name	No.	Name	No.
	Yerkes	26	Picard M	Abbot		Proclus GA	60
	Curtis	27	Picard L	Anville		Proclus V	1
	Picard	28	Da Vinci A	Apollonius L	37	Rima Messier	
	Crile	29	Lick N	Asada		Rimae Apollonius	
	Palus Somni	30	Picard K	Bombelli		Rimae Secchi	
	Glaisher	31	Cauchy V	Cameron		Rimae Taruntius	
	Greaves	32	Cauchy U	Catena Taruntius		Secchi	
	Mare Crisium	33	Lick L	Cauchy D	22	Secchi A	48
	Lick	34	Picard P	Cauchy U	32	Secchi B	46
	Dorsum Termier	35	Taruntius X	Cauchy V	31	Secchi G	45
	Sinus Concordiae	36	Taruntius Z	Cauchy W	19	Secchi K	57
	Da Vinci	37	Apollonius L	Crile		Secchi U	51
	Watts	38	Taruntius R	Curtis		Secchi UA	74
	Tebbutt	39	Taruntius L	Da Vinci		Secchi UB	73
	Shapley	40	Taruntius W	Da Vinci A	28	Secchi X	58
	Lawrence	41	Taruntius U	Dorsa Cato		Shapley	
	Asada	42	Taruntius S	Dorsa Geikie		Sinus Concordiae	
	Cameron	43	Taruntius V	Dorsum Cayeux		Smithson	
	Rimae Taruntius	44	Taruntius F	Dorsum Cushman		Taruntius	
	Zähringer	45	Secchi G	Dorsum Termier		Taruntius B	49
	Taruntius	46	Secchi B	Glaisher		Taruntius CA	66
	Abbot	47	Taruntius T	Glaisher A	9	Taruntius CB	65
	Bombelli	48	Secchi A	Glaisher B	14	Taruntius EB	69
	Rimae Apollonius	49	Taruntius B	Glaisher E	12	Taruntius F	44
	Catena Taruntius	50	Taruntius O	Glaisher F	3	Taruntius H	54
	Montes Secchi	51	Secchi U	Glaisher G	13	Taruntius K	53
	Secchi	52	Lubbock S	Glaisher H	2	Taruntius L	39
	Mare Fecunditatis	53	Taruntius K	Glaisher L	4	Taruntius MA	64
	Anville	54	Taruntius H	Glaisher M	7	Taruntius MB	63
	Smithson	55	Taruntius P	Glaisher N	6	Taruntius O	50
	Mount Marilyn	56	Lubbock R	Glaisher V	17	Taruntius P	55
	Rimae Secchi	57	Secchi K	Glaisher W	11	Taruntius R	38
	Dorsa Cato	58	Secchi X	Greaves		Taruntius S	42
	Dorsum Cushman	59	Messier B	Lawrence		Taruntius T	47
	Dorsum Cayeux	60	Proclus GA	Lick		Taruntius T*	72
	Rima Messier	61	Lick BA	Lick A	15	Taruntius TA	70
	Dorsa Geikie	62	Taruntius η	Lick B	18	Taruntius U	41
1	Proclus V	63	Taruntius MB	Lick BA	61	Taruntius V	43
2	Glaisher H	64	Taruntius MA	Lick C	16	Taruntius VB	71
3	Glaisher F	65	Taruntius CB	Lick E	20	Taruntius W	40
4	Glaisher L	66	Taruntius CA	Lick F	23	Taruntius WA	67
5	Proclus A	67	Taruntius WA	Lick G	24	Taruntius WB	68
6	Glaisher N	68	Taruntius WB	Lick K	25	Taruntius X	35
7	Glaisher M	69	Taruntius EB	Lick L	33	Taruntius Z	36
8	Proclus C	70	Taruntius TA	Lick N	29	Taruntius η	62
9	Glaisher A	71	Taruntius VB	Lubbock R	56	Tebbutt	
10	Proclus G	72	Taruntius T*	Lubbock S	52	Watts	
11	Glaisher W	73	Secchi UB	Mare Crisium		Yerkes	
12	Glaisher E	74	Secchi UA	Mare Fecunditatis		Zähringer	
13	Glaisher G			Messier B	59		
14	Glaisher B			Montes Secchi			
15	Lick A			Mount Marilyn			
16	Lick C			Palus Somni			
17	Glaisher V			Picard			
18	Lick B			Picard K	30		
19	Cauchy W			Picard L	27		
20	Lick E			Picard M	26		
21	Picard N			Picard N	21		
22	Cauchy D			Picard P	34		
23	Lick F			Proclus A	5		
24	Lick G			Proclus C	8		
25	Lick K			Proclus G	10		

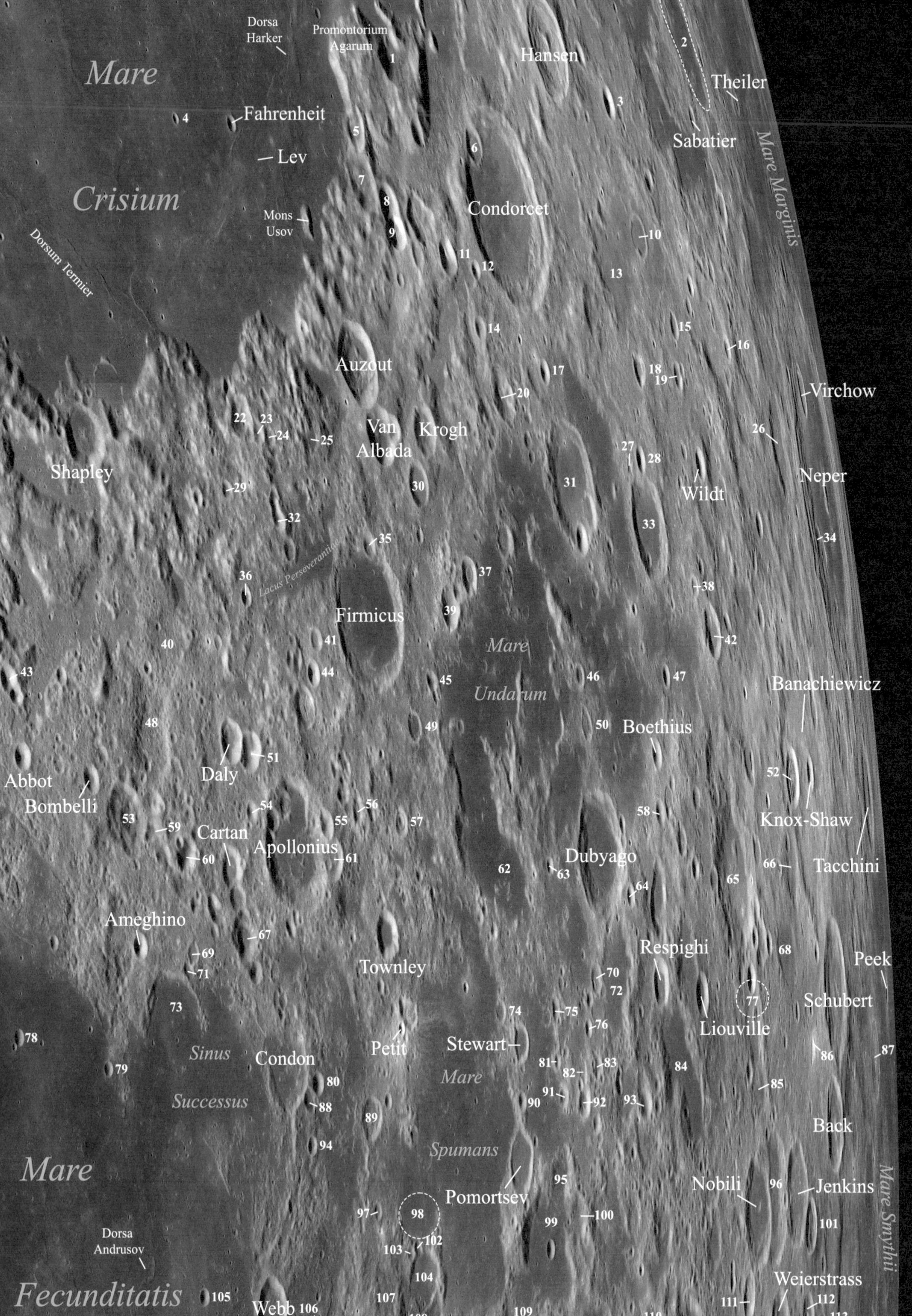
Mare
Crisium
Dorsum Termier
Dorsa Harker
Promontorium Agarum
Fahrenheit
Lev
Mons Usov
Hansen
Theiler
Sabatier
Mare Marginis
Condorcet
Auzout
Van Albada
Krogh
Virchow
Shapley
Neper
Wildt
Lacus Perseverantiae
Firmicus
Mare
Undarum
Banachiewicz
Boethius
Abbot
Bombelli
Daly
Knox-Shaw
Cartan
Apollonius
Dubyago
Tacchini
Ameghino
Respighi
Peek
Townley
Liouville
Schubert
Petit
Stewart
Sinus
Successus
Condon
Mare
Spumans
Back
Mare
Pomortsev
Nobili
Jenkins
Mare Smythii
Dorsa Andrusov
Fecunditatis
Webb
Weierstrass
1
2
3
4
5
6
7
8
9
10
11
12
13
14
15
16
17
18
19
20
22
23
24
25
26
27
28
29
30
31
32
33
34
35
36
37
38
39
40
41
42
43
44
45
46
47
48
49
50
51
52
53
54
55
56
57
58
59
60
61
62
63
64
65
66
67
68
69
70
71
72
73
74
75
76
77
78
79
80
81
82
83
84
85
86
87
88
89
90
91
92
93
94
95
96
97
98
99
100
101
102
103
104
105
106
107
108
109
110
111
112
113

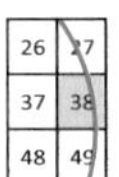

No.	Name	No.	Name	Name	No.	Name	No.
	Dorsa Harker	34	Neper Q	Abbot		Firmicus D	49
	Promontorium Agarum	35	Firmicus E	Ameghino		Firmicus E	35
	Hansen	36	Firmicus H	Apollonius		Firmicus F	44
	Theiler	37	Firmicus C	Apollonius A	53	Firmicus G	41
	Fahrenheit	38	Banachiewicz E	Apollonius B	48	Firmicus H	36
	Lev	39	Firmicus B	Apollonius E	61	Firmicus M	62
	Sabatier	40	Apollonius X	Apollonius F	51	Gilbert P	111
	Mare Marginis	41	Firmicus G	Apollonius H	67	Gilbert V	113
	Mons Usov	42	Banachiewicz C	Apollonius J	59	Gilbert W	112
	Condorcet	43	Apollonius L	Apollonius L	43	Hansen	
	Dorsum Termier	44	Firmicus F	Apollonius M	55	Hansen A	3
	Auzout	45	Firmicus A	Apollonius N	57	Hansen B	2
	Virchow	46	Dubyago W	Apollonius S	89	Jenkins	
	Shapley	47	Dubyago X	Apollonius U	54	Knox-Shaw	
	Van Albada	48	Apollonius B	Apollonius V	60	Krogh	
	Krogh	49	Firmicus D	Apollonius X	40	Lacus Perseverantiae	
	Wildt	50	Dubyago V	Apollonius Y	56	Lev	
	Neper	51	Apollonius F	Auzout		Liouville	
	Lacus Perseverantiae	52	Banachiewicz B	Auzout C	30	Maclaurin H	108
	Firmicus	53	Apollonius A	Auzout D	25	Maclaurin K	109
	Mare Undarum	54	Apollonius U	Auzout E	22	Maclaurin L	110
	Banachiewicz	55	Apollonius M	Auzout L	32	Maclaurin O	99
	Boethius	56	Apollonius Y	Auzout R	29	Maclaurin W	95
	Abbot	57	Apollonius N	Auzout U	23	Maclaurin X	100
	Bombelli	58	Dubyago T	Auzout V	24	Mare Fecunditatis	
	Daly	59	Apollonius J	Back		Mare Marginis	
	Knox-Shaw	60	Apollonius V	Banachiewicz		Mare Smythii	
	Cartan	61	Apollonius E	Banachiewicz B	52	Mare Spumans	
	Apollonius	62	Firmicus M	Banachiewicz C	42	Mare Undarum	
	Dubyago	63	Dubyago Y	Banachiewicz E	38	Mons Usov	
	Tacchini	64	Dubyago Z	Boethius		Neper	
	Ameghino	65	Schubert G	Bombelli		Neper D	26
	Townley	66	Schubert E	Cartan		Neper H	16
	Respighi	67	Apollonius H	Condon		Neper Q	34
	Peek	68	Schubert F	Condorcet		Nobili	
	Schubert	69	Webb X	Condorcet A	11	Peek	
	Petit	70	Dubyago J	Condorcet D	20	Petit	
	Stewart	71	Webb W	Condorcet E	12	Picard Y	4
	Condon	72	Dubyago B	Condorcet F	33	Pomortsev	
	Liouville	73	Webb P	Condorcet G	14	Promontorium Agarum	
	Sinus Successus	74	Dubyago R	Condorcet H	7	Respighi	
	Mare Spumans	75	Dubyago M	Condorcet J	5	Sabatier	
	Back	76	Dubyago H	Condorcet L	18	Schubert	
	Mare Fecunditatis	77	Schubert K	Condorcet M	28	Schubert A	86
	Dorsa Andrusov	78	Taruntius O	Condorcet N	27	Schubert C	87
	Pomortsev	79	Webb U	Condorcet P	31	Schubert E	66
	Nobili	80	Webb G	Condorcet Q	13	Schubert F	68
	Jenkins	81	Dubyago L	Condorcet R	10	Schubert G	65
	Webb	82	Dubyago G	Condorcet S	15	Schubert H	85
	Weierstrass	83	Dubyago F	Condorcet T	9	Schubert J	101
	Mare Smythii	84	Schubert N	Condorcet TA	8	Schubert K	77
1	Condorcet W	85	Schubert H	Condorcet U	19	Schubert N	84
2	Hansen B	86	Schubert A	Condorcet W	1	Schubert X	96
3	Hansen A	87	Schubert C	Condorcet X	17	Shapley	
4	Picard Y	88	Webb F	Condorcet Y	6	Sinus Successus	
5	Condorcet J	89	Apollonius S	Daly		Stewart	
6	Condorcet Y	90	Dubyago N	Dorsa Andrusov		Tacchini	
7	Condorcet H	91	Dubyago K	Dorsa Harker		Taruntius O	78
8	Condorcet TA	92	Dubyago E	Dorsum Termier		Theiler	
9	Condorcet T	93	Dubyago D	Dubyago		Townley	
10	Condorcet R	94	Webb E	Dubyago B	72	Van Albada	
11	Condorcet A	95	Maclaurin W	Dubyago D	93	Virchow	
12	Condorcet E	96	Schubert X	Dubyago E	92	Webb	
13	Condorcet Q	97	Webb L	Dubyago F	83	Webb B	105
14	Condorcet G	98	Webb C	Dubyago G	82	Webb C	98
15	Condorcet S	99	Maclaurin O	Dubyago H	76	Webb E	94
16	Neper H	100	Maclaurin X	Dubyago J	70	Webb F	88
17	Condorcet X	101	Schubert J	Dubyago K	91	Webb G	80
18	Condorcet L	102	Webb M	Dubyago L	81	Webb J	104
19	Condorcet U	103	Webb N	Dubyago M	75	Webb K	107
20	Condorcet D	104	Webb J	Dubyago N	90	Webb L	97
22	Auzout E	105	Webb B	Dubyago R	74	Webb M	102
23	Auzout U	106	Webb Q	Dubyago T	58	Webb N	103
24	Auzout V	107	Webb K	Dubyago V	50	Webb P	73
25	Auzout D	108	Maclaurin H	Dubyago W	46	Webb Q	106
26	Neper D	109	Maclaurin K	Dubyago X	47	Webb U	79
27	Condorcet N	110	Maclaurin L	Dubyago Y	63	Webb W	71
28	Condorcet M	111	Gilbert P	Dubyago Z	64	Webb X	69
29	Auzout R	112	Gilbert W	Fahrenheit		Weierstrass	
30	Auzout C	113	Gilbert V	Firmicus		Wildt	
31	Condorcet P			Firmicus A	45		
32	Auzout L			Firmicus B	39		
33	Condorcet F			Firmicus C	37		

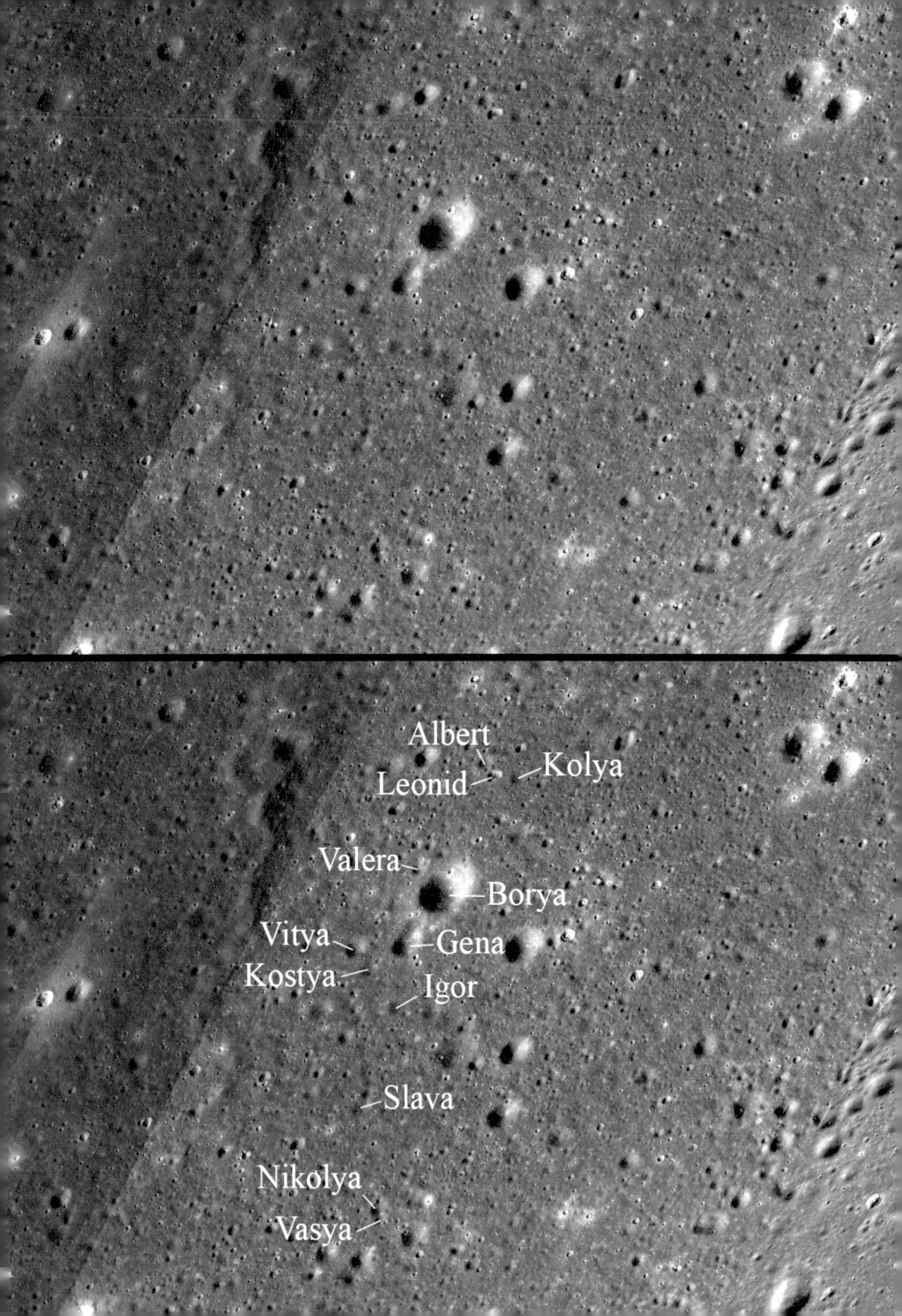

Albert
Kolya
Leonid
Valera
Borya
Vitya
Gena
Kostya
Igor
Slava
Nikolya
Vasya

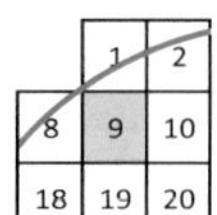

Detail 1

No.	Name	No.	Name	Name	No.	Name	No.
	[illegible]						
	[illegible]						
	Gena						
	Igor						
	Kolya						
	Kostya						
	Leonid						
	Nikolya						
	Slava						
	Valera						
	Vasya						
	Vitya						

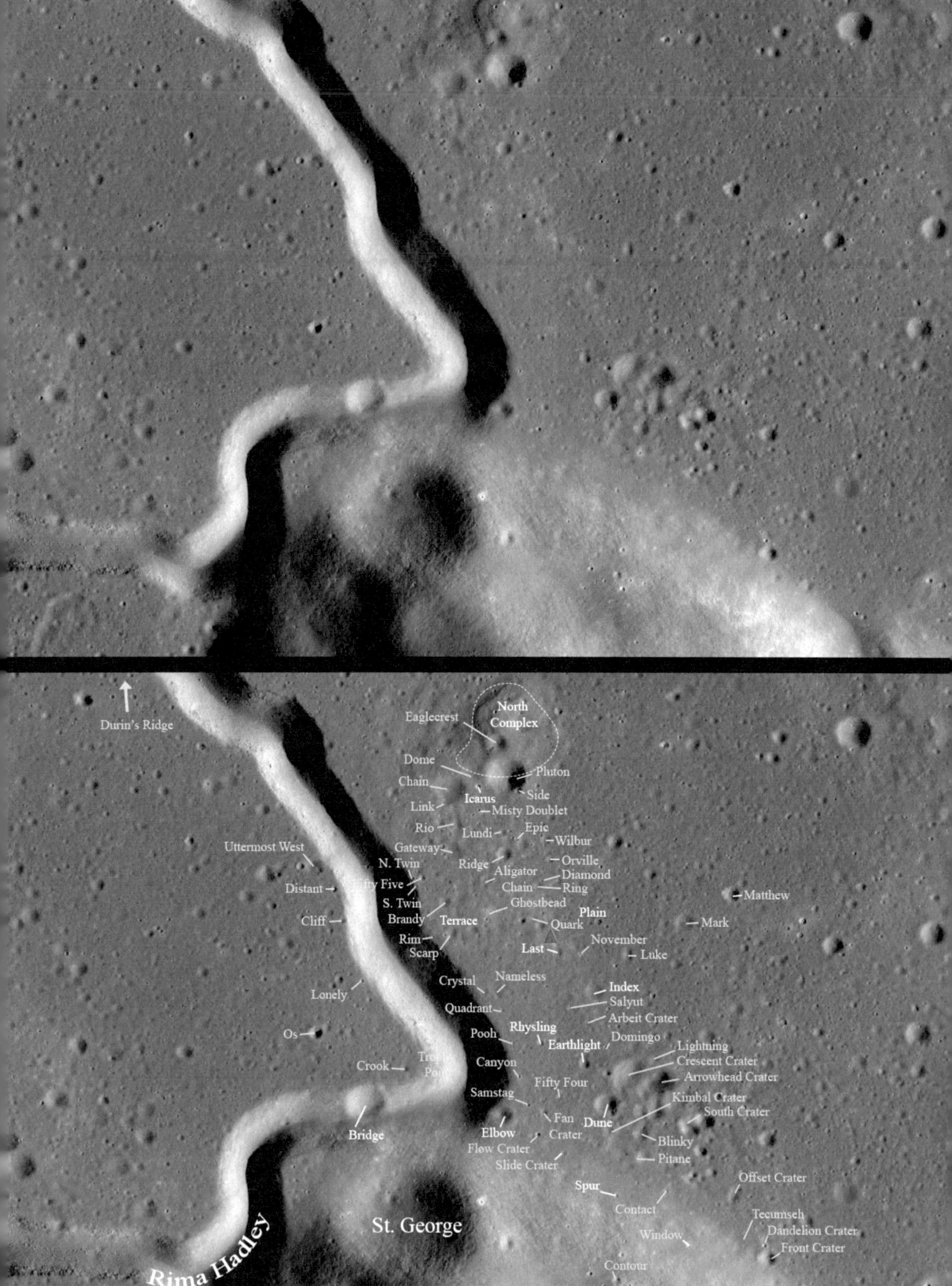
Durin's Ridge
Eaglecrest
North Complex
Dome
Pluton
Chain
Icarus
Side
Link
Misty Doublet
Rio
Lundi
Epic
Wilbur
Uttermost West
Gateway
Ridge
Orville
N. Twin
Aligator
Diamond
Distant
Fifty Five
Chain
Ring
S. Twin
Ghostbead
Matthew
Plain
Cliff
Brandy
Terrace
Quark
Mark
Rim
Last
November
Scarp
Luke
Crystal
Nameless
Lonely
Index
Quadrant
Salyut
Arbeit Crater
Rhysling
Os
Pooh
Earthlight
Domingo
Lightning
Crook
Canyon
Crescent Crater
Arrowhead Crater
Fifty Four
Samstag
Kimbal Crater
South Crater
Fan
Dune
Elbow
Bridge
Crater
Blinky
Flow Crater
Pitane
Slide Crater
Offset Crater
Spur
Contact
Tecumseh
St. George
Dandelion Crater
Window
Rima Hadley
Front Crater
Contour
Apennine Front
Exuperay
High

11	12	13
21	22	23
32	33	34

Detail 2

The red triangle shows the approximate position where the lunar module of the **Apollo 15** mission landed.

No.	Name	No.	Name	Name	No.	Name	No.
	Aligator Chain		Rim				
	Apennine Front		Ring				
	Arbeit Crater		Rio				
	Arrowhead Crater		S. Twin				
	Blinky		Salyut				
	Brandy		Samstag				
	Bridge		Scarp				
	Canyon		Side				
	Chain		Slide Crater				
	Cliff		South Crater				
	Contact		Spur				
	Contour		St. George				
	Crescent Crater		Tecumseh				
	Crook		Terrace				
	Crystal		Trophy Point				
	Dendelion Crater		Uttermost West				
	Diamond		Wilbur				
	Distant		Window				
	Dome						
	Domingo						
	Dune						
	Dune						
	Durin's Ridge						
	Eaglecrest						
	Elbow						
	Epic						
	Exuperay						
	Fan Crater						
	Fifty Five						
	Fifty Four						
	Flow Crater						
	Front Crater						
	Gateway						
	Ghostbead						
	High						
	Icarus						
	Index						
	Kimbal Crater						
	Last						
	Lightning						
	Link						
	Lonely						
	Luke						
	Lundi						
	Mark						
	Matthew						
	Misty Doublet						
	N. Twin						
	Nameless						
	North Complex						
	November						
	Offset Crater						
	Orville						
	Os						
	Pitane						
	Plain						
	Pluton						
	Pooh						
	Quadrant						
	Quark						
	Rhysling						
	Ridge						

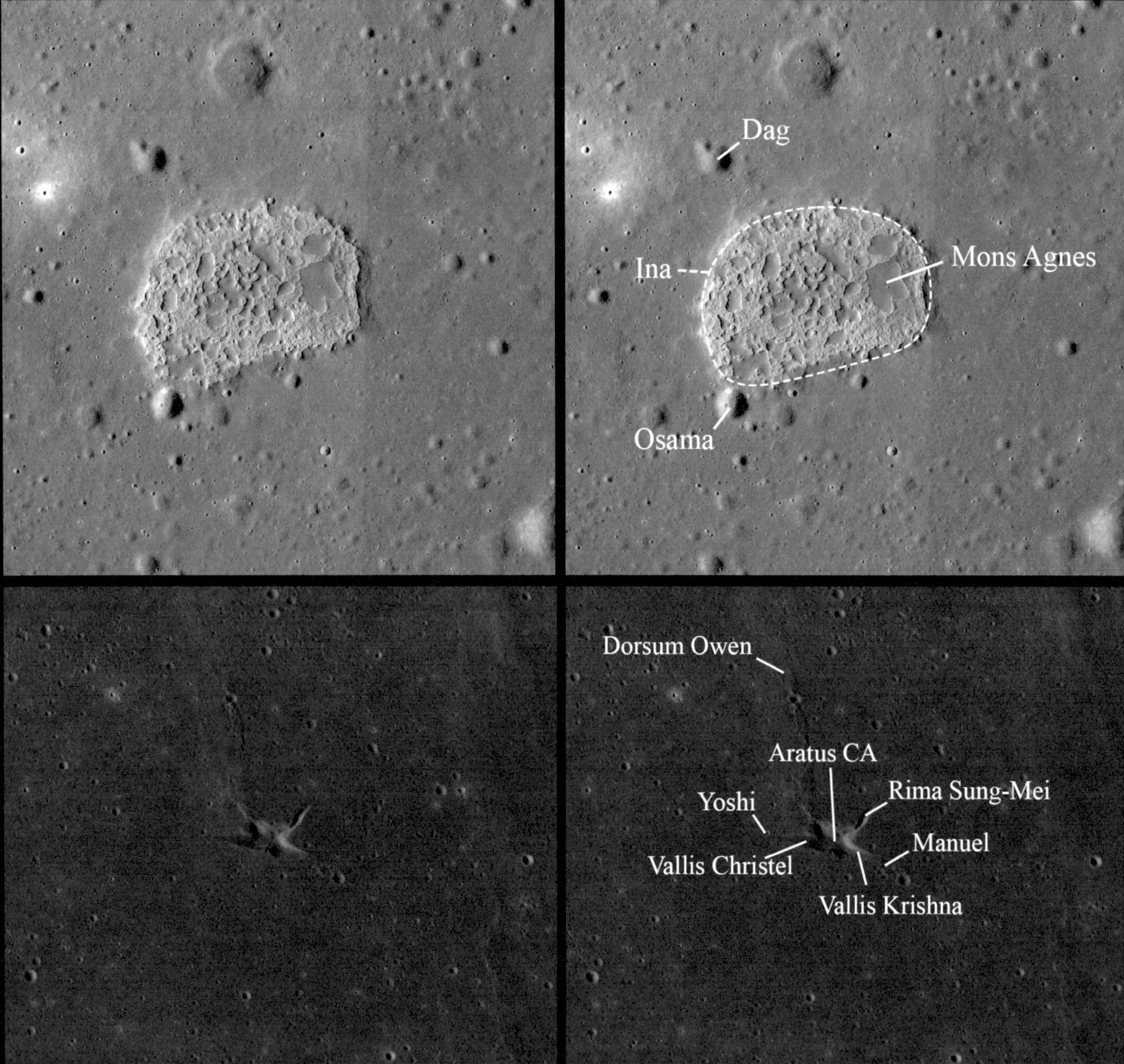
Dag
Ina
Mons Agnes
Osama
Dorsum Owen
Aratus CA
Rima Sung-Mei
Yoshi
Manuel
Vallis Christel
Vallis Krishna

12	13	14
22	23	24
33	34	35

Detail 3+4

No.	Name	No.	Name	Name	No.	Name	No.

3

[illegible]
Ina
Mons Agnes
Osama

4

Aratus CA
Dorsum Owen
Manuel
Rima Sung-Mei
Vallis Christel
Vallis Krishna
Yoshi

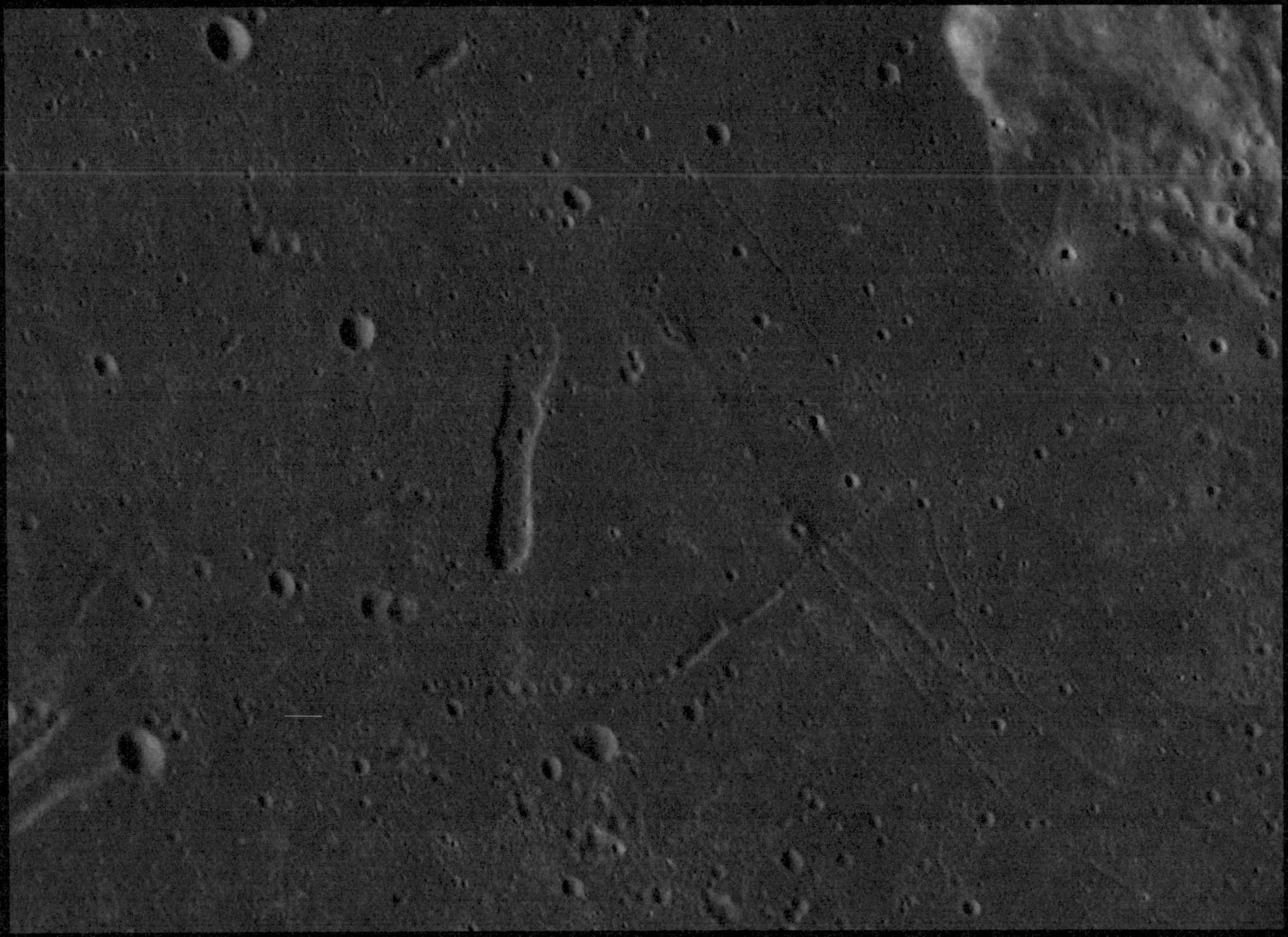

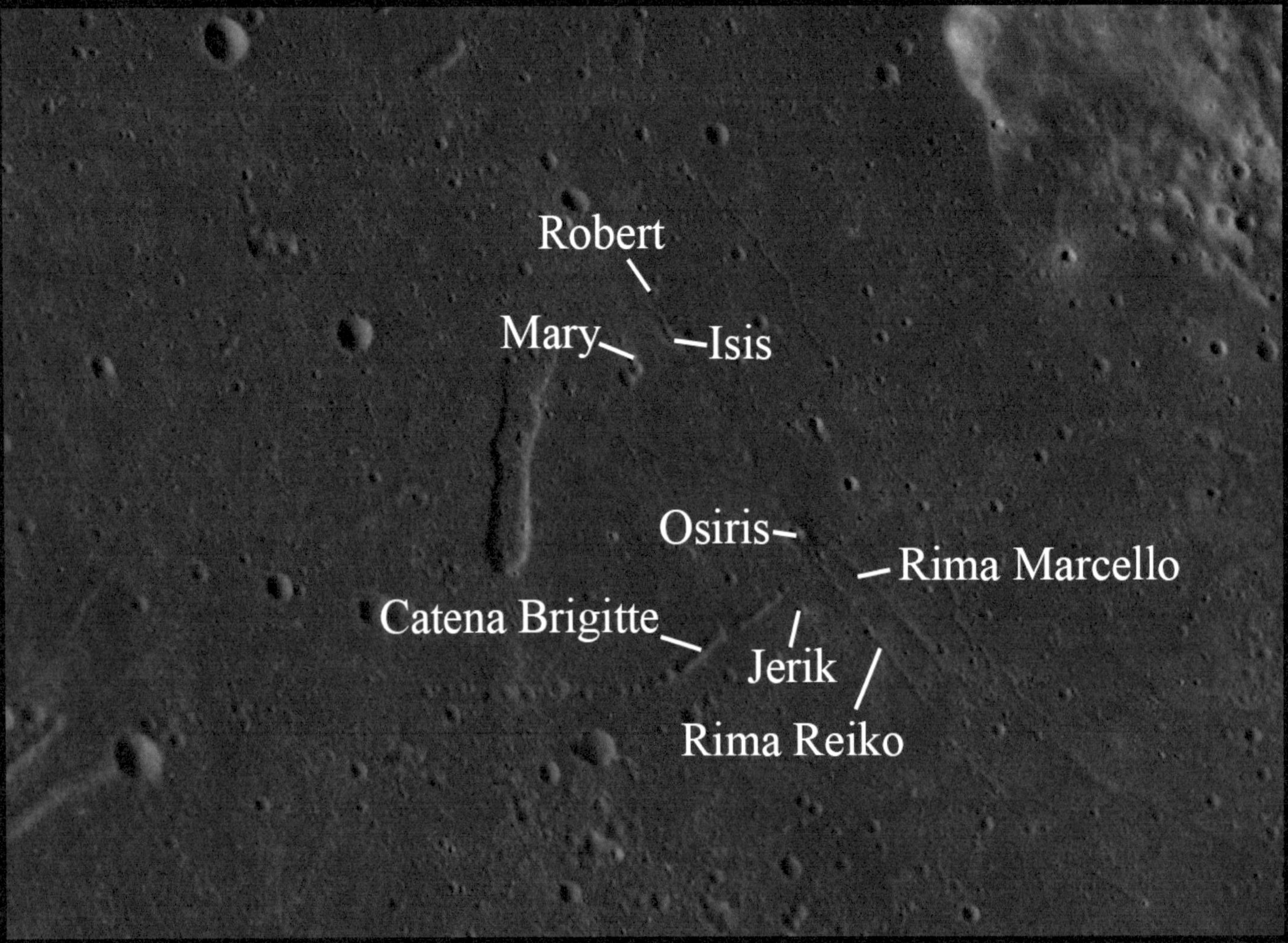

Robert
Mary
Isis
Osiris
Rima Marcello
Catena Brigitte
Jerik
Rima Reiko

13	14	15
23	24	25
34	35	36

Detail 5

No	Name	No.	Name	Name	No.	Name	No.
	Catena Brigitte						
	Isis						
	Jerik						
	Mary						
	Osiris						
	Rima Marcello						
	Rima Reiko						
	Robert						

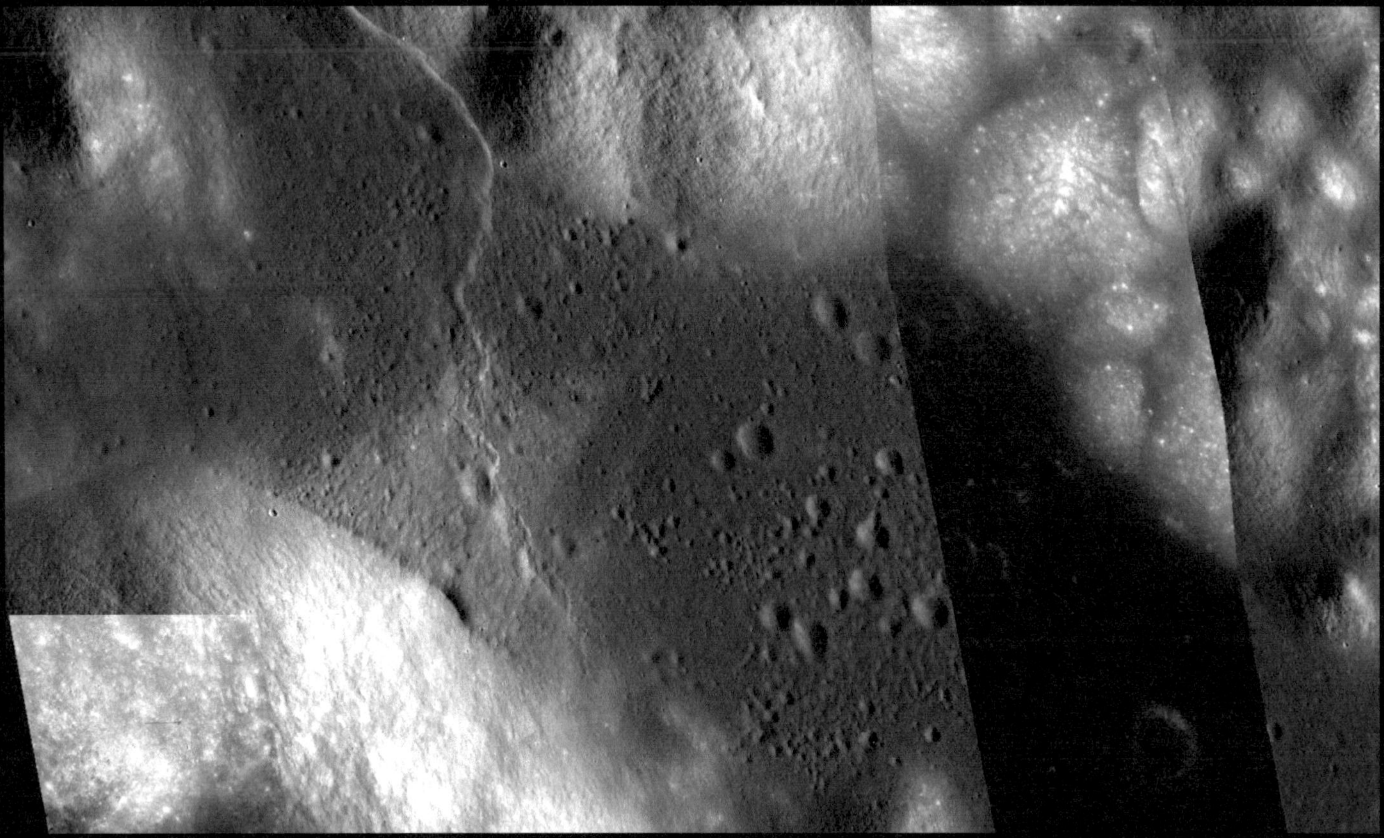

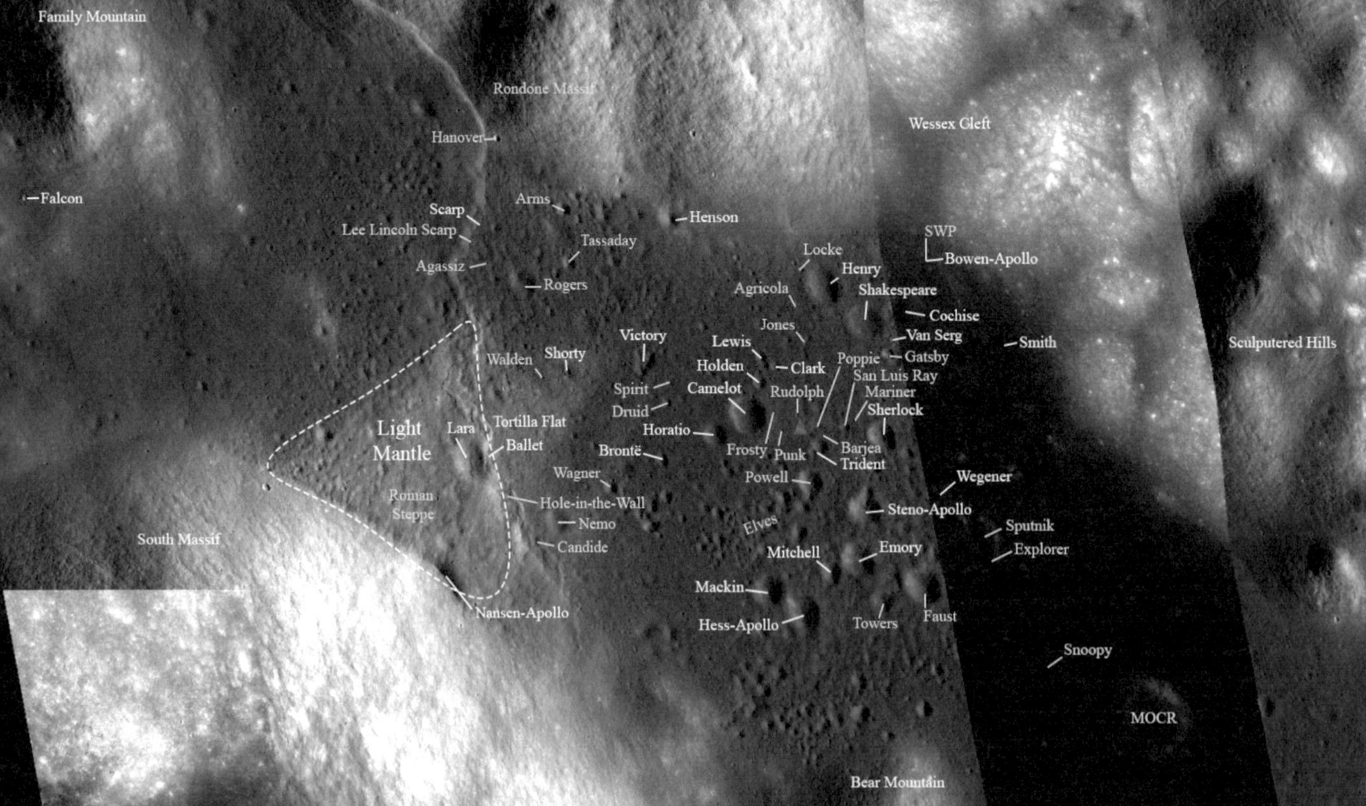

Family Mountain
Rondone Massif
Hanover
Wessex Gleft
Falcon
Scarp
Lee Lincoln Scarp
Arms
Henson
Tassaday
Agassiz
Rogers
Locke
SWP
Bowen-Apollo
Henry
Agricola
Shakespeare
Cochise
Jones
Van Serg
Smith
Sculputered Hills
Victory
Lewis
Poppie
Gatsby
Walden
Shorty
Clark
Holden
San Luis Ray
Spirit
Camelot
Rudolph
Mariner
Druid
Sherlock
Light Mantle
Lara
Tortilla Flat
Horatio
Ballet
Brontë
Frosty
Punk
Barjea
Trident
Wagner
Powell
Wegener
Roman Steppe
Hole-in-the-Wall
Steno-Apollo
Nemo
Elves
Sputnik
South Massif
Candide
Mitchell
Emory
Explorer
Mackin
Nansen-Apollo
Hess-Apollo
Towers
Faust
Snoopy
MOCR
Bear Mountain

14	15	16
24	25	26
35	36	37

Detail 6

No.	Name	No.	Name	Name	No.	Name	No.
	Agassiz		Towers				
	[illegible]		[illegible]				
	Arms		Van Serg				
	Ballet		Victory				
	Barjea		Wagner				
	Bear Mountain		Walden				
	Bowen-Apollo		Wegener				
	Brontë		Wessex Cleft				
	Camelot						
	Candide						
	Clark						
	Cochise						
	Druid						
	Elves						
	Emory						
	Explorer						
	Falcon						
	Family Mountain						
	Faust						
	Frosty						
	Gatsby						
	Hanover						
	Henry						
	Henson						
	Hess-Apollo						
	Holden						
	Hole-in-the-Wall						
	Horatio						
	Jones						
	Lara						
	Lee Lincoln Scarp						
	Lewis						
	Light Mantle						
	Locke						
	Mackin						
	Mariner						
	Mitchell						
	MOCR						
	Nansen-Apollo						
	Nemo						
	Poppie						
	Powel						
	Punk						
	Rogers						
	Roman Steppe						
	Rondone Massif						
	Rudolph						
	San Luis Ray						
	Scarp						
	Sculptured Hills						
	Shakespeare						
	Sherlock						
	Shorty						
	Smith						
	Snoopy						
	South Massif						
	Spirit						
	Sputnik						
	Steno-Apollo						
	SWP						
	Tassaday						
	Tortilla Flat						

The red triangle shows the approximate position where the lunar module of the **Apollo 17** mission landed.

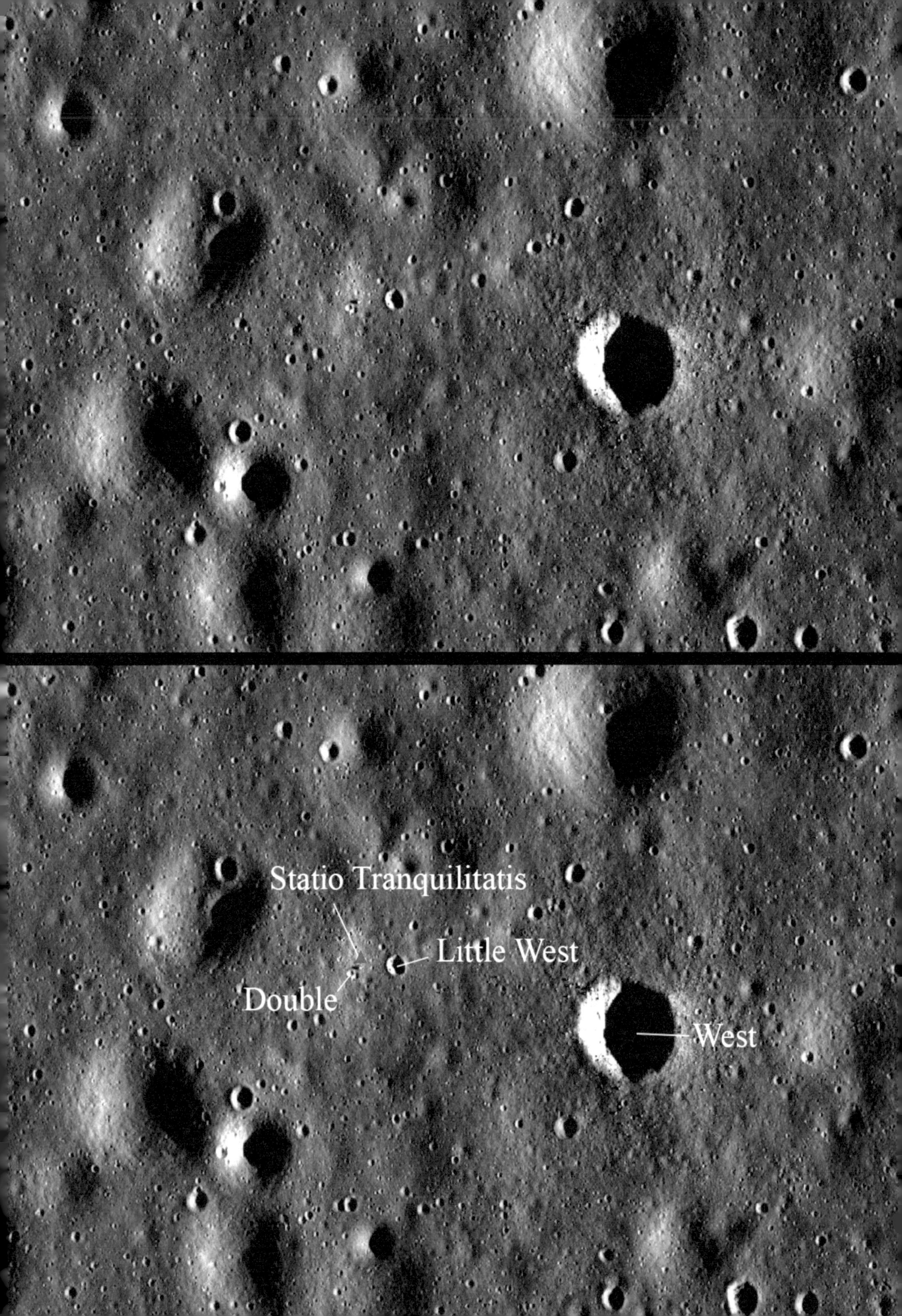
Statio Tranquilitatis
Little West
Double
West

23	24	25
34	35	36
45	46	47

Detail 7

No.	Name	No.	Name	Name	No.	Name	No.
	Double						
	Little West						
	Statio Tranquillitatis						
	West						

The red triangle shows the approximate position where the lunar module of the **Apollo 11** mission landed.

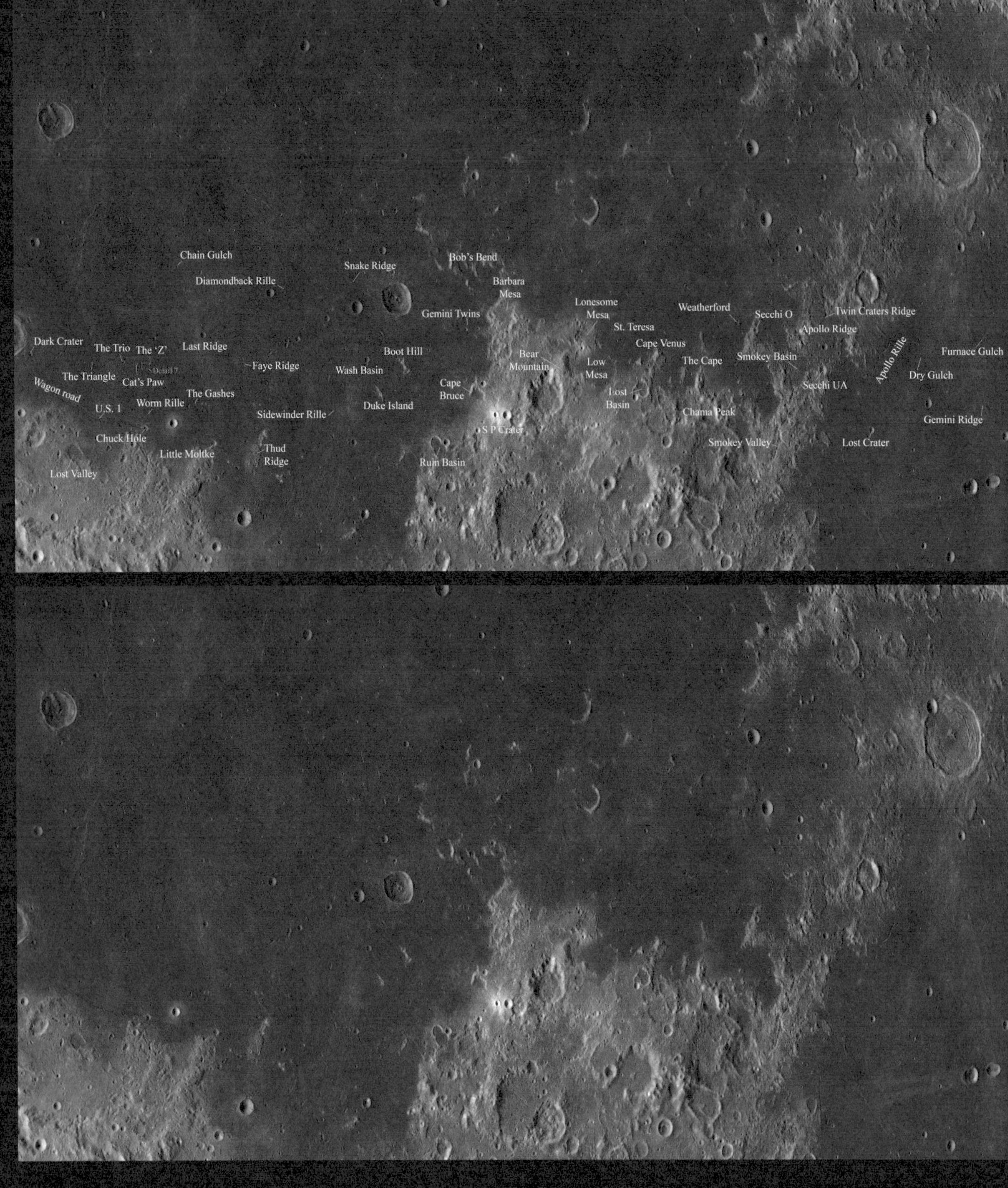

Chain Gulch
Diamondback Rille
Snake Ridge
Bob's Bend
Barbara Mesa
Gemini Twins
Lonesome Mesa
Weatherford
Secchi O
Twin Craters Ridge
St. Teresa
Apollo Ridge
Dark Crater
The Trio
The 'Z'
Last Ridge
Boot Hill
Cape Venus
Furnace Gulch
Apollo Rille
Faye Ridge
Wash Basin
Bear Mountain
Low Mesa
The Cape
Smokey Basin
Dry Gulch
The Triangle
Cat's Paw
Wagon road
Cape Bruce
Lost Basin
Secchi UA
The Gashes
Worm Rille
U.S. 1
Duke Island
Sidewinder Rille
Chama Peak
Gemini Ridge
S P Crater
Chuck Hole
Smokey Valley
Lost Crater
Little Moltke
Thud Ridge
Rum Basin
Lost Valley

23	24	25	26	27
34	35	36	37	38
45	46	47	48	49

Detail 8

No.	Name	No.	Name	Name	No.	Name	No.
	Apollo Ridge						
	Apollo Rille						
	Barbara Mesa						
	Bear Mountain						
	Bob's Bend						
	Boot Hill						
	Cahin Gulch						
	Cape Bruce						
	Cape Venus						
	Cat's Paw						
	Chama Peak						
	Chuck Hole						
	Dark Crater						
	Diamondback Rille						
	Dry Gulch						
	Duke Island						
	Faye Ridge						
	Furnace Gulch						
	Gemini Ridge						
	Last Ridge						
	Little Moltke						
	Lonesome Mesa						
	Lost Basin						
	Lost Crater						
	Lost Valley						
	Low Mesa						
	Ruin Basin						
	S P Crater						
	Secchi O						
	Secchi UA						
	Sidewinder Rille						
	Smokey Basin						
	Smokey Valley						
	Snake Ridge						
	St. Teresa						
	The Cape						
	The Gashes						
	The Triangle						
	The Trio						
	The 'Z'						
	Thud Ridge						
	Twin Craters Ridge						
	U.S. 1						
	Wagon Road						
	Wash Basin						
	Weatherford						
	Worm Rille						

End Notes

Feature	Comment
Scoresby AA	Scoresby AA is situated on the rim of De Sitter A. However, in the A&G De Sitter A is named as Scoresby A. Hence the use of the main feature name Scoresby.
Louville B	The A&G shows Louville B to be the northern crater of the two craters in Louville. IAU recognises the southern crater as Louville B.
Helicon C*	NASA's LAC 24 shows Helicon C to be a different Crater, marked with number 17a in this map.
Posidonius Z	Atlas and Gazetteer of the near side of the Moon shows this as Posidonius WA in map 86-1, also LAC 42.
Posidonius Y	Atlas and Gazetteer of the near side of the Moon shows this as Posidonius γ in map 86-1, also LAC 42.
Aristarchus δ	A&G also notes Herodotus δ at the same location (A&G 158-1). Since Herodotus δ is also annotated under number 54, it is assumed this is Aristarchus δ.
T. Mayer K	A&G Notes this as T. Mayer GA on map 133-2
Le Monnier KA	In Atlas and Gazetteer this feature is named KA and listed under Posidonius. NASA's LAC 42 shows the name Le Monnier KA. Given the absence of Posidonius K and the presence of nearby Le Monnier K and KB, I conclude that the listing in the Atlas and Gazetteer must be wrong and Le Monnier KA should be the actual correct name.
Linné ED	Comparing Atlas and Gazetteer with NASA LAC-42 the position of ED seems to be different on both.
Jansen EB	Two different Craters were given the same name on different maps. The northern EB "option" was noted on Map 78-2, whilst the southern EB "option" was noted on 85-2, both in the Atlas and Gazetteer.
Römer KA	Crater Franck used to be called Römer K, hence the KA
Littrow BA	Crater Clerke used to be called Littrow B, hence the BA
Littrow BC	Crater Clerke used to be called Littrow B, hence the BC
Macrobius BA	Crater Hill used to be called Macrobius B, hence the BA
Macrobius AA	Crater Carmichael used to be called Macrobius A, hence the AA. NLC on Map LAC43 shows the position of Macrobius AA to be the larger crater on the east of the group of three to be Macrobius AA. Here indicated with the dashed line."
Macrobius AB	Crater Carmichael used to be called Macrobius A, hence the AB
Yerkes V	This feature was not named in the A&G
Reiner R	Ghost Crater
Lansberg ϕ	AIC 58D shows this to be Kunowsky ϕ
Gay-Lussac H	Atlas and Gazetteer of the near side of the Moon shows the wrong crater in their map 126-2 for H.
Reinhold δ	The elevation Reinhold Delta is drawn into LAC58 at a different position than it is noted on map 126-1 of the Atlas and Gazetteer. The location of the LAC58 is noted with the straight line whilst the position of the Atlas and Gazetteer is noted with a dotted line.
Donna	The elevation Donna is designated as Cauchy ω in A&G maps 66-1 & 73-1
Cauchy V	Currently this crater is unnamed. In A&G map 66-2, this feature has the name Cauchy V, The current Cauchy V is named Cauchy U and the current Cauchy U is named Cauchy J. Cauchy J is currently unused
Taruntius S	Formerly known as Taruntius EA
Taruntius T*	Taruntius T and TB seem to have swapped names: Formerly the smaller, deeper crater was TB, now it is designated T. The larger crater is currently not named and was formerly known as T. (A&G 61-1, AIC 61C)
Icarus	Icarus is also a feature on the IAU list with coordinate of -5,49 Lat, 186,74 Long.
Scarp	Scarp is also a named feature in the Taurus-Littrow Valley; that one seems to be included in the official list of the IAU.

Index

The index section is built up using an alphabetised list of all features. The alphabetisation has been done strictly according to the exact spelling used in the atlas (e.g. G. Bond is under "G" and Rima Jansen is under "R").

Apart from the feature name, the index provides following information:

- **Source** The source of the feature name, which can be one of the following:
  - → IAU International Astronomical Union; this is an officially named feature.
  - → A&G Atlas and Gazetteer of the Near Side of the Moon, published by NASA
  - → AIC Apollo Intermediate Charts, published by NASA
  - → CAF Coordinates of Anthropogenic Features on the Moon
  - → LC Luna Cognita
  - → NLC Lunar Chart Published for the USAF and NASA, 1st Edition 1963, also called LAC, just like the current IAU maps.
- **Reference** Within the specified source, where can the feature be found. This is either a page or a map reference. E.g. for all IAU approved features, an LAC map number will be cited and for all A&G features the number will be that of which map the feature can be found on.
- **Map(s)** On which maps in this atlas can the feature be found. The feature may appear on more than one map; either because it is large, or because it is situated on one of the overlap areas of two maps.

In the below example the feature named Taruntius TA is originally found in the Atlas and Gazetteer of the Near Side of the Moon, on map 61-1. It is found on map 37 in this atlas.

Feature	Source	Reference	Map(s)
Taruntius TA	A&G	61-1	37

Feature	Source	Reference	Map(s)
Abbot	IAU	LAC62	37, 38
Abetti	IAU	LAC42	24
Agassiz	CAF	B6	Detail 6
Agricola	CAF	B6	Detail 6
Agrippa	IAU	LAC60	34
Agrippa B	IAU	LAC59	34
Agrippa D	IAU	LAC59	34
Agrippa E	IAU	LAC59	34
Agrippa F	IAU	LAC60	34
Agrippa FA	AIC	60D	34
Agrippa G	IAU	LAC59	34
Agrippa H	IAU	LAC60	34
Agrippa S	IAU	LAC59	34
Akis	IAU	LAC39	19
Al-Bakri	IAU	LAC60	24, 35
Albert	IAU	LAC24	Detail 1
Aldrin	IAU	LAC60	35
Alexander	IAU	LAC26	13
Alexander A	IAU	LAC26	13
Alexander B	IAU	LAC26	13
Alexander C	IAU	LAC26	13
Alexander K	IAU	LAC26	13
Alexander β	A&G	98-2	13
Alexander κ	A&G	98-2	13
Alexander ϕ	A&G	98-2	13
Alexander ψ	A&G	98-2	13
Alhazen	IAU	LAC45	27
Alhazen A	IAU	LAC45	27
Alhazen D	IAU	LAC45	27
Alhazen α	NLC	LAC44	27
Alhazen β	NLC	LAC44	27
Aligator Chain	CAF	B4	Detail 2
Aloha	IAU	LAC38	8, 18
Alpes A	IAU	LAC12	4
Alpes AB	A&G	116-1	4
Alpes B	IAU	LAC25	12
Ameghino	IAU	LAC62	38
Anaxagoras	IAU	LAC3	4
Anaxagoras A	IAU	LAC3	4
Anaxagoras B	IAU	LAC3	4
Anaximander	IAU	LAC2	2
Anaximander A	IAU	LAC2	2
Anaximander B	IAU	LAC2	2
Anaximander D	IAU	LAC2	2
Anaximander H	IAU	LAC2	2, 3
Anaximander R	IAU	LAC2	2
Anaximander S	IAU	LAC2	2
Anaximander T	IAU	LAC2	2
Anaximander U	IAU	LAC2	2
Anaximenes	IAU	LAC2	3
Anaximenes B	IAU	LAC2	3
Anaximenes E	IAU	LAC3	3
Anaximenes G	IAU	LAC2	3
Anaximenes H	IAU	LAC2	3
Anaximenes HA	A&G	164-2	3

Feature	Source	Reference	Map(s)
Anaximenes HB	A&G	164-2	3
Ango	IAU	LAC39	19
Angström	IAU	LAC39	9, 19
Angström A	IAU	LAC39	9
Angström B	IAU	LAC39	9
Ann	IAU	LAC41	22
Annegrit	IAU	LAC40	20, 22
Anville	IAU	LAC61	37
Apennine Front	IAU	LAC41	Detail 2
Apollo Ridge	CAF	B1	Detail 8
Apollo Rille	CAF	B1	Detail 8
Apollonius	IAU	LAC62	38
Apollonius A	IAU	LAC62	38
Apollonius B	IAU	LAC62	38
Apollonius E	IAU	LAC62	38
Apollonius F	IAU	LAC62	38
Apollonius H	IAU	LAC62	38
Apollonius J	IAU	LAC62	38
Apollonius L	IAU	LAC62	37, 38
Apollonius M	IAU	LAC62	38
Apollonius N	IAU	LAC62	38
Apollonius S	IAU	LAC62	38
Apollonius U	IAU	LAC62	38
Apollonius V	IAU	LAC62	38
Apollonius X	IAU	LAC62	38
Apollonius Y	IAU	LAC62	38
Arago	IAU	LAC60	35
Arago B	IAU	LAC60	35
Arago C	IAU	LAC60	35
Arago CA	AIC	60C	35
Arago D	IAU	LAC60	35
Arago E	IAU	LAC60	35
Arago EA	AIC	60C	35
Arago α	AIC	60C	35
Arago β	AIC	60D	35
Aratus	IAU	LAC41	22
Aratus B	IAU	LAC41	22, 23
Aratus C	IAU	LAC41	23
Aratus CA	IAU	LAC42	23, Detail 4
Aratus CC	A&G	97-3	23
Aratus D	IAU	LAC41	23
Aratus θ	A&G	102-3	22
Arbeit Crater	CAF	B4	Detail 2
Archimedes	IAU	LAC41	12, 22
Archimedes AA	A&G	115-1	21
Archimedes AB	A&G	115-1	21
Archimedes C	IAU	LAC41	12
Archimedes D	IAU	LAC25	12
Archimedes E	IAU	LAC41	21
Archimedes G	IAU	LAC41	21
Archimedes H	IAU	LAC41	21
Archimedes L	IAU	LAC41	22
Archimedes M	IAU	LAC41	22
Archimedes N	IAU	LAC41	22
Archimedes P	IAU	LAC41	22

Feature	Source	Reference	Map(s)
Archimedes Q	IAU	LAC41	22
Archimedes R	IAU	LAC41	21
Archimedes S	IAU	LAC41	12, 22
Archimedes T	IAU	LAC41	12
Archimedes U	IAU	LAC25	12
Archimedes V	IAU	LAC25	12
Archimedes W	IAU	LAC41	21
Archimedes X	IAU	LAC41	11
Archimedes Y	IAU	LAC41	11, 21
Archimedes Z	IAU	LAC41	22
Archimedes γ	A&G	109-3	22
Archimedes δ	A&G	109-3	22
Archimedes ϵ	A&G	110-2	12
Archimedes ζ	A&G	115-1	11
Archimedes μ	A&G	115-1	22
Archimedes ξ	A&G	115-1	12
Archimedes ϕ	A&G	109-3	22
Archytas	IAU	LAC12	4
Archytas B	IAU	LAC12	4
Archytas C	A&G	116-1	4
Archytas D	IAU	LAC13	5
Archytas DA	NLC	LAC13	5
Archytas G	IAU	LAC12	4
Archytas K	IAU	LAC12	4
Archytas L	IAU	LAC12	4
Archytas U	IAU	LAC12	4
Archytas W	IAU	LAC12	4
Ariadaeus	IAU	LAC60	35
Ariadaeus A	IAU	LAC60	35
Ariadaeus B	IAU	LAC60	34
Ariadaeus BA	A&G	90-1	34
Ariadaeus D	IAU	LAC60	35
Ariadaeus DA	A&G	90-1	34, 35
Ariadaeus E	IAU	LAC60	35
Ariadaeus F	IAU	LAC60	35
Ariadaeus α	AIC	60D	34
Ariadaeus ϵ	AIC	60D	34, 35
Aristarchus	IAU	LAC39	18
Aristarchus B	IAU	LAC39	18
Aristarchus D	IAU	LAC39	19
Aristarchus F	IAU	LAC39	18
Aristarchus H	IAU	LAC39	18
Aristarchus N	IAU	LAC39	18, 19
Aristarchus S	IAU	LAC39	18
Aristarchus T	IAU	LAC39	18
Aristarchus U	IAU	LAC39	18
Aristarchus Z	IAU	LAC39	18
Aristarchus δ	A&G	158-1	18
Aristarchus μ	A&G	158-1	18
Aristillus	IAU	LAC25	12
Aristillus A	IAU	LAC25	12
Aristillus B	IAU	LAC25	12
Aristoteles	IAU	LAC13	5
Aristoteles D	IAU	LAC26	5, 13
Aristoteles M	IAU	LAC13	5, 6
Aristoteles N	IAU	LAC13	5, 6
Aristoteles γ	A&G	103-3	[illegible]
Aristoteles ζ	A&G	103-3	5
Aristoteles θ	A&G	103-3	5
Arms	CAF	B6	Detail 6
Armstrong	IAU	LAC60	35
Arnold	IAU	LAC4	5
Arnold A	IAU	LAC4	5
Arnold E	IAU	LAC4	5
Arnold F	IAU	LAC4	5
Arnold G	IAU	LAC4	5
Arnold H	IAU	LAC4	5
Arnold J	IAU	LAC4	5
Arnold K	IAU	LAC4	5
Arnold L	IAU	LAC4	5
Arnold M	IAU	LAC4	5
Arnold N	IAU	LAC4	5
Arnold κ	A&G	92-2	5
Arrowhead Crater	CAF	B4	Detail 2
Artemis	IAU	LAC40	20
Artsimovich	IAU	LAC39	19
Aryabhata	IAU	LAC61	36
Asada	IAU	LAC61	37
Aston	IAU	LAC36	8
Aston K	IAU	LAC36	8
Aston L	IAU	LAC36	8
Atlas	IAU	LAC27	7, 15
Atlas A	IAU	LAC27	15
Atlas AA	A&G	74-2	15
Atlas D	IAU	LAC14	7
Atlas E	IAU	LAC14	6, 7, 14, 15
Atlas G	IAU	LAC14	6, 7
Atlas L	IAU	LAC14	7
Atlas P	IAU	LAC14	7
Atlas W	IAU	LAC27	15
Atlas X	IAU	LAC27	15
Atlas ϵ	A&G	67-3	7
Autolycus	IAU	LAC41	12
Autolycus A	IAU	LAC41	12
Autolycus B	A&G	109-3	22
Autolycus K	IAU	LAC41	12
Autolycus α	A&G	109-3	22
Autolycus β	A&G	102-3	22
Autolycus γ	A&G	102-3	22
Autolycus η	A&G	110-1	12
Autolycus ω	A&G	110-1	12
Auwers	IAU	LAC60	24
Auwers A	IAU	LAC60	24, 35
Auzout	IAU	LAC62	38
Auzout C	IAU	LAC62	38
Auzout D	IAU	LAC62	38
Auzout E	IAU	LAC62	38
Auzout L	IAU	LAC62	38
Auzout R	IAU	LAC62	38
Auzout U	IAU	LAC62	38

Feature	Source	Reference	Map(s)
Auzout V	IAU	LAC62	38
Babbage	IAU	LAC10	1, 2
Babbage A	IAU	LAC10	2
Babbage B	IAU	LAC10	1
Babbage C	IAU	LAC10	2
Babbage D	IAU	LAC10	1, 2
Babbage E	IAU	LAC10	1, 2
Babbage U	IAU	LAC10	2
Babbage X	IAU	LAC10	2
Back	IAU	LAC63	38
Baillaud	IAU	LAC4	5
Baillaud A	IAU	LAC4	5
Baillaud B	IAU	LAC4	5
Baillaud C	IAU	LAC4	5
Baillaud D	IAU	LAC4	5
Baillaud E	IAU	LAC4	5
Baillaud F	IAU	LAC4	5
Baily	IAU	LAC13	6
Baily A	IAU	LAC13	6, 14
Baily B	IAU	LAC13	6
Baily K	IAU	LAC13	6
Balboa	IAU	LAC37	17
Balboa A	IAU	LAC37	17
Balboa B	IAU	LAC37	17
Balboa C	IAU	LAC37	17
Balboa D	IAU	LAC37	17
Ballet	IAU	LAC43	Detail 6
Banachiewicz	IAU	LAC63	38
Banachiewicz B	IAU	LAC63	38
Banachiewicz C	IAU	LAC63	38
Banachiewicz E	IAU	LAC63	38
Bancroft	IAU	LAC41	21, 22
Banting	IAU	LAC42	23
Barbara Mesa	CAF	B1	Detail 8
Barjea	CAF	B6	Detail 6
Barrow	IAU	LAC3	4
Barrow A	IAU	LAC3	4
Barrow B	IAU	LAC4	4
Barrow C	IAU	LAC4	4
Barrow E	IAU	LAC3	4
Barrow F	IAU	LAC3	4
Barrow G	IAU	LAC3	4
Barrow H	IAU	LAC3	4
Barrow K	IAU	LAC4	4
Barrow KA	A&G	116-2	4
Barrow KB	A&G	116-2	4
Barrow M	IAU	LAC3	4
Beals	IAU	LAC29	16
Bear Mountain	IAU	LAC43	Detail 6, Detail 8
Beer	IAU	LAC41	21
Beer A	IAU	LAC41	21
Beer B	IAU	LAC41	21
Beer E	IAU	LAC41	21
Beketov	IAU	LAC42	25
Béla	IAU	LAC41	22
Bernoulli	IAU	LAC28	16
Bernoulli A	IAU	LAC28	16
Bernoulli B	IAU	LAC28	16
Bernoulli C	IAU	LAC28	16
Bernoulli D	IAU	LAC28	16
Bernoulli E	IAU	LAC28	16
Bernoulli K	IAU	LAC28	16
Berosus	IAU	LAC28	16
Berosus A	IAU	LAC28	16
Berosus F	IAU	LAC28	16
Berosus K	IAU	LAC28	16
Berzelius	IAU	LAC27	15
Berzelius A	IAU	LAC27	15
Berzelius B	IAU	LAC27	15
Berzelius F	IAU	LAC27	15
Berzelius FA	A&G	74-1	15
Berzelius FB	A&G	74-1	15
Berzelius K	IAU	LAC27	15
Berzelius T	IAU	LAC27	15
Berzelius W	IAU	LAC27	15
Bessarion	IAU	LAC57	19
Bessarion A	IAU	LAC39	19
Bessarion B	IAU	LAC39	18, 19
Bessarion C	IAU	LAC39	18
Bessarion D	IAU	LAC39	18, 19
Bessarion E	IAU	LAC57	19
Bessarion G	IAU	LAC57	18, 19
Bessarion H	IAU	LAC57	18, 19
Bessarion V	IAU	LAC57	19
Bessarion W	IAU	LAC39	19
Bessarion ζ	A&G	138-2	19, 30
Bessarion η	A&G	133-2	19, 30
Bessarion θ	A&G	138-2	19, 30
Bessarion λ	A&G	133-2	19, 30
Bessarion ξ	A&G	133-2	19, 30
Bessel	IAU	LAC42	24
Bessel AB	A&G	91-1	24
Bessel AC	A&G	91-1	24
Bessel AD	A&G	91-1	24
Bessel D	IAU	LAC42	24
Bessel DA	A&G	91-1	24
Bessel F	IAU	LAC42	23
Bessel FA	A&G	90-3	23
Bessel FB	A&G	90-3	23
Bessel FC	A&G	90-3	23
Bessel G	IAU	LAC42	23
Bessel GA	A&G	90-3	23
Bessel GB	A&G	90-3	23
Bessel H	IAU	LAC42	24
Bianchini	IAU	LAC11	2, 10
Bianchini A	A&G	152-1	2
Bianchini D	IAU	LAC24	2, 10
Bianchini G	IAU	LAC24	10
Bianchini H	IAU	LAC11	2, 10
Bianchini M	IAU	LAC11	2, 10

Feature	Source	Reference	Map(s)
Bianchini N	IAU	LAC11	2, 10
Bianchini [illegible]	A&G	[illegible]	[illegible]
Bianchini W	IAU	LAC11	2, 10
Birmingham	IAU	LAC3	3, 4
Birmingham B	IAU	LAC12	3, 4
Birmingham G	IAU	LAC3	4
Birmingham H	IAU	LAC3	4
Birmingham K	IAU	LAC3	3, 4
Birmingham β	A&G	128-4	3, 4
Birmingham ϵ	A&G	128-3	3
Birmingham ζ	A&G	128-1	3
Birmingham η	A&G	140-2	3
Birmingham λ	A&G	128-6	3, 4
Birmingham μ	A&G	128-5	3
Birmingham ξ	A&G	128-2	3
Blagg	IAU	LAC59	33
Blinky	CAF	B4	Detail 2
Bliss	IAU	LAC12	3
Bobillier	IAU	LAC42	23
Bob's Bend	CAF	B1	Detail 8
Bode	IAU	LAC59	33
Bode A	IAU	LAC59	33
Bode B	IAU	LAC59	33
Bode BA	A&G	109-2	33
Bode C	IAU	LAC59	33
Bode D	IAU	LAC59	33
Bode E	IAU	LAC59	33
Bode EA	A&G	109-2	33
Bode G	IAU	LAC59	33
Bode H	IAU	LAC59	32
Bode K	IAU	LAC59	33
Bode L	IAU	LAC59	33
Bode N	IAU	LAC59	33
Boethius	IAU	LAC63	38
Bohr	IAU	LAC55	28
Bombelli	IAU	LAC62	37, 38
Boole	IAU	LAC21	2
Boole A	IAU	LAC10	1, 2
Boole B	IAU	LAC10	2
Boole C	IAU	LAC9	2
Boole D	IAU	LAC21	2
Boole E	IAU	LAC21	2
Boole F	IAU	LAC10	2
Boole R	IAU	LAC2	2
Boot Hill	CAF	B1	Detail 8
Borel	IAU	LAC42	24
Boris	IAU	LAC39	9
Borya	IAU	LAC24	Detail 1
Boscovich	IAU	LAC60	34
Boscovich A	IAU	LAC60	34
Boscovich B	IAU	LAC59	34
Boscovich C	IAU	LAC60	34
Boscovich D	IAU	LAC60	34
Boscovich E	IAU	LAC60	34
Boscovich F	IAU	LAC60	34
Boscovich P	IAU	LAC60	34
[illegible]	[illegible]	[illegible]	[illegible]
Boss B	IAU	LAC15	7
Boss C	IAU	LAC15	7
Boss K	IAU	LAC15	7
Boss L	IAU	LAC15	7
Bouguer	IAU	LAC11	2
Bouguer A	IAU	LAC11	2
Bouguer B	IAU	LAC11	2
Bowen	IAU	LAC41	23
Bowen-Apollo	IAU	LAC43	Detail 6
Brackett	IAU	LAC42	24
Bradley H	IAU	LAC41	22
Bradley K	IAU	LAC41	22
Bradley ϕ	A&G	109-2	22
Brandy	CAF	B4	Detail 2
Brayley	IAU	LAC39	19
Brayley B	IAU	LAC39	19
Brayley C	IAU	LAC39	19
Brayley D	IAU	LAC39	19
Brayley E	IAU	LAC39	19
Brayley F	IAU	LAC39	19
Brayley G	IAU	LAC39	19
Brayley K	IAU	LAC39	19
Brayley L	IAU	LAC39	18, 19
Brayley S	IAU	LAC39	19
Brayley α	A&G	138-3	19
Brayley π	A&G	138-3	19
Brayley σ	A&G	144-3	19
Brewster	IAU	LAC43	25
Bridge	IAU	LAC41	Detail 2
Briggs	IAU	LAC38	17
Briggs A	IAU	LAC37	17
Briggs B	IAU	LAC37	17
Briggs C	IAU	LAC38	17
Briggs η	A&G	175-1	8, 17
Briggs ξ	A&G	169-3	17
Brontë	IAU	LAC43	Detail 6
Bruce	IAU	LAC59	33
Bunsen	IAU	LAC36	8
Bunsen D	IAU	LAC36	8
Burckhardt	IAU	LAC44	16
Burckhardt A	IAU	LAC44	16
Burckhardt B	IAU	LAC44	16, 26
Burckhardt C	IAU	LAC44	16
Burckhardt E	IAU	LAC44	16
Burckhardt F	IAU	LAC44	16
Burckhardt FA	A&G	62-1	16
Burckhardt G	IAU	LAC27	16
Bürg	IAU	LAC26	14
Bürg A	IAU	LAC26	14
Bürg B	IAU	LAC26	14
Byrd	IAU	LAC1	4
C. Herschel	IAU	LAC24	10
C. Herschel C	IAU	LAC24	10

Feature	Source	Reference	Map(s)
C. Herschel E	IAU	LAC24	9
C. Herschel U	IAU	LAC24	10
C. Herschel V	IAU	LAC24	9, 10
C. Herschel ϵ	A&G	145-1	9
C. Herschel ζ	A&G	145-1	9, 10
C. Mayer	IAU	LAC13	5
C. Mayer B	IAU	LAC13	5
C. Mayer D	IAU	LAC13	5
C. Mayer E	IAU	LAC13	5
C. Mayer F	IAU	LAC13	5
C. Mayer H	IAU	LAC13	5
Cahin Gulch	CAF	B1	Detail 8
Cajal	IAU	LAC61	36
Calippus	IAU	LAC26	13
Calippus A	IAU	LAC25	13
Calippus B	IAU	LAC26	13
Calippus C	IAU	LAC25	13
Calippus D	IAU	LAC26	13
Calippus E	IAU	LAC26	13
Calippus F	IAU	LAC25	13
Calippus G	IAU	LAC26	13
Calippus α	A&G	103-1	13
Calippus β	A&G	103-1	13
Calippus γ	A&G	103-1	13
Calippus λ	A&G	103-1	13
Calippus ν	A&G	103-1	13
Camelot	IAU	LAC43	Detail 6
Cameron	IAU	LAC61	37
Candide	CAF	B6	Detail 6
Cannon	IAU	LAC45	27
Cannon B	IAU	LAC45	27
Cannon E	IAU	LAC45	27
Canyon	CAF	B4	Detail 2
Cape Bruce	CAF	B1	Detail 8
Cape Venus	CAF	B1	Detail 8
Cardanus	IAU	LAC55	28
Cardanus B	IAU	LAC55	28
Cardanus C	IAU	LAC55	28
Cardanus E	IAU	LAC55	28
Cardanus G	IAU	LAC55	28
Cardanus K	IAU	LAC55	17, 28
Cardanus M	IAU	LAC55	17
Cardanus R	IAU	LAC55	28
Carlini	IAU	LAC24	10
Carlini A	IAU	LAC24	10
Carlini C	IAU	LAC24	10
Carlini D	IAU	LAC24	11
Carlini DA	A&G	122-1	11
Carlini DB	A&G	122-1	11
Carlini E	IAU	LAC40	10
Carlini G	IAU	LAC24	10
Carlini H	IAU	LAC24	10
Carlini K	IAU	LAC40	10
Carlini L	IAU	LAC40	10
Carlini S	IAU	LAC24	10

Feature	Source	Reference	Map(s)
Carlos	IAU	LAC41	22
Carmichael	IAU	LAC43	25
Carpenter	IAU	LAC2	2, 3
Carpenter T	IAU	LAC2	2
Carpenter U	IAU	LAC2	2
Carpenter V	IAU	LAC2	3
Carpenter W	IAU	LAC2	2, 3
Carpenter Y	IAU	LAC2	2
Carrel	IAU	LAC60	35
Carrington	IAU	LAC28	15, 16
Cartan	IAU	LAC62	38
Cassini	IAU	LAC25	12
Cassini A	IAU	LAC25	12
Cassini B	IAU	LAC25	12
Cassini C	IAU	LAC25	12, 13
Cassini E	IAU	LAC25	12, 13
Cassini F	IAU	LAC25	12, 13
Cassini G	IAU	LAC25	12
Cassini K	IAU	LAC25	12
Cassini L	IAU	LAC25	12
Cassini M	IAU	LAC25	12
Cassini N	A&G	110-2	12
Cassini P	IAU	LAC25	12
Cassini W	IAU	LAC25	12
Cassini X	IAU	LAC25	12, 13
Cassini Y	IAU	LAC25	12
Cassini Z	IAU	LAC25	12
Catena Brigitte	IAU	LAC42	Detail 5
Catena Krafft	IAU	LAC55	28
Catena Littrow	IAU	LAC42	24, 25
Catena Pierre	IAU	LAC39	19
Catena Taruntius	IAU	LAC61	37
Catena Timocharis	IAU	LAC40	21
Catena Yuri	IAU	LAC39	19
Cat's Paw	CAF	B1	Detail 8
Cauchy	IAU	LAC61	36
Cauchy A	IAU	LAC61	36
Cauchy B	IAU	LAC61	36
Cauchy C	IAU	LAC61	36
Cauchy D	IAU	LAC61	36, 37
Cauchy E	IAU	LAC61	36
Cauchy F	IAU	LAC61	36
Cauchy M	IAU	LAC61	36
Cauchy U	IAU	LAC61	37
Cauchy V	A&G	66-2	36, 37
Cauchy W	IAU	LAC61	37
Cauchy ρ	A&G	73-1	36
Cauchy τ	A&G	73-2	36
Cavalerius	IAU	LAC56	28
Cavalerius A	IAU	LAC56	28
Cavalerius B	IAU	LAC55	28
Cavalerius C	IAU	LAC56	28
Cavalerius D	IAU	LAC56	28
Cavalerius E	IAU	LAC55	28
Cavalerius F	IAU	LAC56	28

Feature	Source	Reference	Map(s)
Cavalerius K	IAU	LAC56	28
Cavalerius L	IAU	LAC56	28
Cavalerius M	IAU	LAC55	28
Cavalerius U	IAU	LAC56	28
Cavalerius W	IAU	LAC56	28
Cavalerius X	IAU	LAC56	28
Cavalerius Y	IAU	LAC56	28
Cavalerius Z	IAU	LAC56	28
Caventou	IAU	LAC40	10, 20
Cayley	IAU	LAC60	34
Censorinus	IAU	LAC79	36
Censorinus A	IAU	LAC79	36
Censorinus AB	A&G	73-1	36
Censorinus CA	A&G	73-1	36
Censorinus J	IAU	LAC79	36
Censorinus K	IAU	LAC79	36
Censorinus V	IAU	LAC79	36
Censorinus W	IAU	LAC79	36
Censorinus X	IAU	LAC79	36
Cepheus	IAU	LAC27	15
Cepheus A	IAU	LAC27	15
Chacornac	IAU	LAC43	14, 15, 24, 25
Chacornac A	IAU	LAC43	14, 15, 24, 25
Chacornac B	IAU	LAC43	24, 25
Chacornac BA	A&G	79-1	24, 25
Chacornac C	IAU	LAC43	14, 15
Chacornac D	IAU	LAC43	15
Chacornac E	IAU	LAC43	15, 25
Chacornac F	IAU	LAC43	25
Chain	CAF	B4	Detail 2
Challis	IAU	LAC3	4
Challis A	IAU	LAC3	4
Chama Peak	CAF	B1	Detail 8
Charles	IAU	LAC40	20
Chevallier	IAU	LAC27	15
Chevallier B	IAU	LAC27	15
Chevallier F	IAU	LAC27	15
Chevallier K	IAU	LAC27	15
Chevallier M	IAU	LAC27	15
Ching-Te	IAU	LAC42	25
Chladni	IAU	LAC59	33
Chuck Hole	CAF	B1	Detail 8
Clark	IAU	B6	Detail 6
Cleomedes	IAU	LAC44	26
Cleomedes A	IAU	LAC44	26
Cleomedes B	IAU	LAC44	26
Cleomedes C	IAU	LAC44	26
Cleomedes D	IAU	LAC44	26
Cleomedes DA	A&G	54-3	26
Cleomedes DB	A&G	54-3	26
Cleomedes DC	A&G	55-1	16
Cleomedes DE	A&G	55-1	16, 26
Cleomedes DF	A&G	55-1	16
Cleomedes DG	A&G	55-1	16
Cleomedes E	IAU	LAC44	26

Feature	Source	Reference	Map(s)
Cleomedes F	IAU	LAC44	26
Cleomedes G	IAU	LAC44	26
Cleomedes GA	A&G	54-3	26
Cleomedes H	IAU	LAC44	26
Cleomedes J	IAU	LAC44	26
Cleomedes K	A&G	54-3	26
Cleomedes L	IAU	LAC44	26
Cleomedes M	IAU	LAC44	26
Cleomedes N	IAU	LAC44	26
Cleomedes P	IAU	LAC44	26
Cleomedes Q	IAU	LAC44	26
Cleomedes R	IAU	LAC44	16, 26
Cleomedes S	IAU	LAC44	26
Cleomedes T	IAU	LAC44	26
Cleomedes α	A&G	54-3	26
Cleomedes β	NLC	LAC44	26
Cleostratus	IAU	LAC10	1
Cleostratus A	IAU	LAC10	1, 2
Cleostratus E	IAU	LAC10	1
Cleostratus F	IAU	LAC10	1
Cleostratus H	IAU	LAC21	1
Cleostratus J	IAU	LAC21	1
Cleostratus L	IAU	LAC10	1, 2
Cleostratus M	IAU	LAC10	1, 2
Cleostratus N	IAU	LAC10	1
Cleostratus P	IAU	LAC10	1
Cleostratus R	IAU	LAC10	1
Clerke	IAU	LAC42	24, 25
Cliff	CAF	B4	Detail 2
Cochise	IAU	LAC43	Detail 6
Collins	IAU	LAC60	35
Condon	IAU	LAC62	38
Condorcet	IAU	LAC62	38
Condorcet A	IAU	LAC62	38
Condorcet D	IAU	LAC62	38
Condorcet E	IAU	LAC62	38
Condorcet F	IAU	LAC63	38
Condorcet G	IAU	LAC62	38
Condorcet H	IAU	LAC62	38
Condorcet J	IAU	LAC62	38
Condorcet L	IAU	LAC63	38
Condorcet M	IAU	LAC63	38
Condorcet N	IAU	LAC63	38
Condorcet P	IAU	LAC63	38
Condorcet Q	IAU	LAC63	38
Condorcet R	IAU	LAC63	38
Condorcet S	IAU	LAC63	38
Condorcet T	IAU	LAC62	38
Condorcet TA	IAU	LAC62	38
Condorcet U	IAU	LAC63	38
Condorcet W	IAU	LAC62	27, 38
Condorcet X	IAU	LAC62	38
Condorcet Y	IAU	LAC62	38
Conon	IAU	LAC41	22
Conon A	IAU	LAC41	22

Feature	Source	Reference	Map(s)
Conon W	IAU	LAC41	22
Conon Y	IAU	LAC41	22
Conon Z	A&G	102-3	22
Conon α	A&G	102-3	22
Conon β	A&G	102-3	22
Conon λ	A&G	102-3	22
Conon ϕ	A&G	102-2	22
Contact	CAF	B4	Detail 2
Contour	CAF	B4	Detail 2
Copernicus	IAU	LAC58	31
Copernicus A	IAU	LAC58	31
Copernicus B	IAU	LAC58	31
Copernicus BB	A&G	126-2	31
Copernicus BC	A&G	126-2	31
Copernicus BD	A&G	126-2	31
Copernicus C	IAU	LAC58	31, 32
Copernicus CA	A&G	126-2	31
Copernicus CB	A&G	114-1	32
Copernicus CC	A&G	114-2	32
Copernicus CD	A&G	114-1	32
Copernicus D	IAU	LAC58	31
Copernicus DA	A&G	126-2	31
Copernicus E	IAU	LAC58	31
Copernicus F	IAU	LAC58	31
Copernicus G	IAU	LAC58	31
Copernicus GA	A&G	126-1	31
Copernicus H	IAU	LAC58	31
Copernicus J	IAU	LAC58	31
Copernicus JC	A&G	126-2	31
Copernicus JD	A&G	126-2	31
Copernicus JE	A&G	126-2	31
Copernicus K	A&G	126-2	31
Copernicus KA	A&G	126-2	31
Copernicus L	IAU	LAC58	31
Copernicus N	IAU	LAC58	31
Copernicus P	IAU	LAC58	31, 32
Copernicus PA	A&G	126-2	31
Copernicus R	IAU	LAC58	31
Courtney	IAU	LAC39	19
Crescent Crater	CAF	B4	Detail 2
Crile	IAU	LAC61	26, 37
Crook	CAF	B4	Detail 2
Crystal	CAF	B4	Detail 2
Curtis	IAU	LAC62	26, 27, 37
Cusanus	IAU	LAC5	6
Cusanus A	IAU	LAC5	6
Cusanus B	IAU	LAC5	6
Cusanus C	IAU	LAC5	6
Cusanus E	IAU	LAC5	6
Cusanus F	IAU	LAC5	6
Cusanus H	IAU	LAC5	6
Da Vinci	IAU	LAC61	37
Da Vinci A	IAU	LAC61	37
Dag	IAU	LAC41	Detail 3
Dalton	IAU	LAC37	17

Feature	Source	Reference	Map(s)
Daly	IAU	LAC62	38
Daniell	IAU	LAC26	14
Daniell D	IAU	LAC26	14
Daniell W	IAU	LAC26	14
Daniell X	IAU	LAC26	14
Daniell η	A&G	86-2	14
Daniell ξ	A&G	86-2	14
Dark Crater	CAF	B1	Detail 8
D'Arrest	IAU	LAC60	34
D'Arrest A	IAU	LAC60	34
D'Arrest B	IAU	LAC60	34
D'Arrest BA	A&G	90-1	34
D'Arrest BC	A&G	90-1	34
D'Arrest M	IAU	LAC60	34
D'Arrest R	IAU	LAC60	34
Daubrée	IAU	LAC60	23
Dawes	IAU	LAC42	24
De La Rue	IAU	LAC14	6
De La Rue D	IAU	LAC14	6
De La Rue E	IAU	LAC14	6
De La Rue EA	A&G	68-1	6
De La Rue J	IAU	LAC14	6
De La Rue P	IAU	LAC14	6
De La Rue Q	IAU	LAC14	6
De La Rue R	IAU	LAC14	6
De La Rue S	IAU	LAC14	6
De La Rue W	IAU	LAC14	6
De Morgan	IAU	LAC60	34
De Sitter	IAU	LAC4	5
De Sitter A	IAU	LAC1	4
De Sitter F	IAU	LAC1	5
De Sitter G	IAU	LAC4	5
De Sitter L	IAU	LAC4	5
De Sitter M	IAU	LAC1	4, 5
De Sitter U	IAU	LAC4	5
De Sitter V	IAU	LAC5	5
De Sitter W	IAU	LAC4	5
De Sitter X	IAU	LAC1	5
Debes	IAU	LAC44	16, 26
Debes A	IAU	LAC44	26
Debes B	IAU	LAC44	26
Dechen	IAU	LAC22	8, 9
Dechen A	IAU	LAC22	8, 9
Dechen B	IAU	LAC22	8
Dechen C	IAU	LAC22	8
Dechen D	IAU	LAC23	9
Delambre D	IAU	LAC78	35
Delambre F	IAU	LAC78	35
Delambre FA	A&G	85-1	35
Delambre H	IAU	LAC78	34, 35
Delambre J	IAU	LAC78	35
Delisle	IAU	LAC39	9, 19
Delisle K	IAU	LAC39	19
Delisle α	A&G	145-1	9
Delisle γ	A&G	145-1	9

Feature	Source	Reference	Map(s)
Delisle ϵ	A&G	144-3	19
Delmotte	IAU	LAC44	26
[illegible]	[illegible]	[illegible]	[illegible]
Dembowski A	IAU	LAC59	34
Dembowski B	IAU	LAC59	34
Dembowski C	IAU	LAC59	34
Democritus	IAU	LAC13	5
Democritus A	IAU	LAC13	5
Democritus B	IAU	LAC13	5
Democritus BA	A&G	92-1	5
Democritus D	IAU	LAC13	5
Democritus K	IAU	LAC14	6
Democritus L	IAU	LAC13	5, 6
Democritus M	IAU	LAC13	5
Democritus N	IAU	LAC13	5
Dendelion Crater	CAF	B4	Detail 2
Desargues	IAU	LAC2	2
Desargues A	IAU	LAC2	2
Desargues B	IAU	LAC2	2
Desargues C	IAU	LAC2	2
Desargues D	IAU	LAC2	2
Desargues E	IAU	LAC2	2
Desargues K	IAU	LAC2	2
Desargues L	IAU	LAC9	2
Desargues M	IAU	LAC2	2
Deseilligny	IAU	LAC42	24
Diamond	CAF	B4	Detail 2
Diamondback Rille	CAF	B1	Detail 8
Diana	IAU	LAC61	25, 36
Dionysius	IAU	LAC60	35
Dionysius A	IAU	LAC60	35
Dionysius AB	AIC	60D	35
Dionysius AC	AIC	60D	35
Dionysius B	IAU	LAC60	34, 35
Dionysius β	AIC	60D	35
Dionysius γ	AIC	60D	34, 35
Diophantus	IAU	LAC39	19
Diophantus B	IAU	LAC39	19
Diophantus C	IAU	LAC39	19
Diophantus D	IAU	LAC39	19
Distant	CAF	B4	Detail 2
Dome	CAF	B4	Detail 2
Domingo	CAF	B4	Detail 2
Dominicus Maria	LC	14-36	31
Donna	IAU	LAC61	36
Dorsa Aldrovandi	IAU	LAC42	24
Dorsa Andrusov	IAU	LAC80	38
Dorsa Argand	IAU	LAC39	19
Dorsa Barlow	IAU	LAC61	25
Dorsa Burnet	IAU	LAC38	18
Dorsa Cato	IAU	LAC61	37
Dorsa Geikie	IAU	LAC80	37
Dorsa Harker	IAU	LAC62	27, 38
Dorsa Lister	IAU	LAC42	24
Dorsa Smirnov	IAU	LAC42	24
Dorsa Sorby	IAU	LAC42	23
Dorsa Stille	IAU	LAC10	20
Dorsa Tetyaev	IAU	LAC44	27
Dorsa Whiston	IAU	LAC38	8, 18
Dorsum Arduino	IAU	LAC39	19
Dorsum Azara	IAU	LAC42	24
Dorsum Bucher	IAU	LAC39	9
Dorsum Buckland	IAU	LAC42	23, 24
Dorsum Cayeux	IAU	LAC62	37
Dorsum Cushman	IAU	LAC61	37
Dorsum Gast	IAU	LAC41	23
Dorsum Grabau	IAU	LAC40	21
Dorsum Heim	IAU	LAC24	10
Dorsum Higazy	IAU	LAC40	20, 21
Dorsum Nicol	IAU	LAC42	24
Dorsum Niggli	IAU	LAC38	18
Dorsum Oppel	IAU	LAC44	26
Dorsum Owen	IAU	LAC42	Detail 4
Dorsum Scilla	IAU	LAC38	8
Dorsum Termier	IAU	LAC62	37, 38
Dorsum Thera	IAU	LAC39	19
Dorsum von Cotta	IAU	LAC42	23
Dorsum Zirkel	IAU	LAC40	10, 20
Double	IAU	LAC60	Detail 7
Draper	IAU	LAC40	20
Draper A	IAU	LAC40	20
Draper C	IAU	LAC40	20
Druid	CAF	B6	Detail 6
Dry Gulch	CAF	B1	Detail 8
Dubyago	IAU	LAC62	38
Dubyago B	IAU	LAC63	38
Dubyago D	IAU	LAC63	38
Dubyago E	IAU	LAC62	38
Dubyago F	IAU	LAC62	38
Dubyago G	IAU	LAC62	38
Dubyago H	IAU	LAC62	38
Dubyago J	IAU	LAC62	38
Dubyago K	IAU	LAC62	38
Dubyago L	IAU	LAC62	38
Dubyago M	IAU	LAC62	38
Dubyago N	IAU	LAC62	38
Dubyago R	IAU	LAC62	38
Dubyago T	IAU	LAC63	38
Dubyago V	IAU	LAC62	38
Dubyago W	IAU	LAC62	38
Dubyago X	IAU	LAC63	38
Dubyago Y	IAU	LAC62	38
Dubyago Z	IAU	LAC63	38
Duke Island	CAF	B1	Detail 8
Dune	IAU	LAC41	Detail 2
Durin's Ridge	CAF	B4	Detail 2
Eaglecrest	CAF	B4	Detail 2
Eckert	IAU	LAC44	27
Eddington	IAU	LAC37	17
Eddington P	IAU	LAC37	17

Feature	Source	Reference	Map(s)
Eddington α	A&G	175-1	17
Egede	IAU	LAC13	5, 13
Egede A	IAU	LAC13	5
Egede B	IAU	LAC12	4, 5
Egede C	IAU	LAC13	5
Egede E	IAU	LAC13	5
Egede F	IAU	LAC13	5
Egede G	IAU	LAC12	4
Egede M	IAU	LAC13	5
Egede N	IAU	LAC13	5
Egede P	IAU	LAC26	5, 13
Eimmart	IAU	LAC44	26, 27
Eimmart A	IAU	LAC44	27
Eimmart AB	NLC	LAC44	27
Eimmart B	IAU	LAC44	27
Eimmart C	IAU	LAC44	26, 27
Eimmart CA	A&G	54-3	26
Eimmart D	IAU	LAC44	27
Eimmart DA	NLC	LAC44	27
Eimmart F	IAU	LAC44	26, 27
Eimmart FA	A&G	54-3	26, 27
Eimmart G	IAU	LAC44	26, 27
Eimmart GA	NLC	LAC44	26
Eimmart GB	A&G	54-3	26
Eimmart H	IAU	LAC44	27
Eimmart HA	NLC	LAC44	26, 27
Eimmart K	IAU	LAC44	27
Eimmart KA	NLC	LAC44	27
Eimmart T	NLC	LAC44	26, 27
Eimmart TA	NLC	LAC44	26, 27
Eimmart TB	NLC	LAC44	26
Elbow	IAU	LAC41	Detail 2
Elves	CAF	B6	Detail 6
Emory	IAU	LAC43	Detail 6
Encke	IAU	LAC57	30
Encke B	IAU	LAC57	30
Encke C	IAU	LAC57	30
Encke E	IAU	LAC57	29, 30
Encke G	IAU	LAC57	30
Encke GA	A&G	138-1	30
Encke H	IAU	LAC57	30
Encke J	IAU	LAC57	29, 30
Encke K	IAU	LAC57	30
Encke M	IAU	LAC57	30
Encke N	IAU	LAC57	30
Encke T	IAU	LAC57	29, 30
Encke X	IAU	LAC57	29
Encke Y	IAU	LAC57	30
Encke β	A&G	138-1	29, 30
Encke γ	A&G	138-1	30
Encke θ	A&G	138-1	30
Encke ρ	A&G	138-1	30
Encke σ	A&G	138-1	30
Endymion	IAU	LAC14	7
Endymion A	IAU	LAC14	7

Feature	Source	Reference	Map(s)
Endymion B	IAU	LAC14	6, 7
Endymion BA	A&G	62-3	7
Endymion BB	A&G	62-3	7
Endymion C	IAU	LAC14	6, 7
Endymion CA	A&G	62-3	7
Endymion CB	A&G	62-3	7
Endymion D	IAU	LAC14	7
Endymion E	IAU	LAC14	7
Endymion F	IAU	LAC14	7
Endymion G	IAU	LAC14	6, 7
Endymion H	IAU	LAC14	7
Endymion J	IAU	LAC14	6, 7
Endymion K	IAU	LAC14	7
Endymion L	IAU	LAC15	7
Endymion M	IAU	LAC15	7
Endymion N	IAU	LAC14	7
Endymion W	IAU	LAC14	7
Endymion X	IAU	LAC14	6, 7
Endymion Y	IAU	LAC14	7
Epic	CAF	B4	Detail 2
Epigenes	IAU	LAC3	4
Epigenes A	IAU	LAC3	4
Epigenes B	IAU	LAC3	4
Epigenes D	IAU	LAC3	4
Epigenes F	IAU	LAC3	4
Epigenes G	IAU	LAC3	4
Epigenes H	IAU	LAC3	4
Epigenes P	IAU	LAC3	4
Eratosthenes	IAU	LAC58	21, 32
Eratosthenes A	IAU	LAC41	21
Eratosthenes B	IAU	LAC41	21
Eratosthenes C	IAU	LAC40	21
Eratosthenes D	IAU	LAC40	21
Eratosthenes E	IAU	LAC40	21
Eratosthenes F	IAU	LAC41	21
Eratosthenes G	A&G	114-2	21
Eratosthenes H	IAU	LAC58	32
Eratosthenes K	IAU	LAC59	32
Eratosthenes KA	A&G	114-2	32
Eratosthenes KB	A&G	114-2	32
Eratosthenes M	IAU	LAC58	21, 32
Eratosthenes Z	IAU	LAC58	21, 32
Esclangon	IAU	LAC43	25
Euctemon	IAU	LAC4	5
Euctemon C	IAU	LAC4	5
Euctemon D	IAU	LAC4	5
Euctemon H	IAU	LAC4	5
Euctemon K	IAU	LAC4	5
Euctemon N	IAU	LAC4	5
Eudoxus	IAU	LAC26	13
Eudoxus A	IAU	LAC26	13
Eudoxus B	IAU	LAC26	13
Eudoxus D	IAU	LAC26	13
Eudoxus E	IAU	LAC26	13
Eudoxus G	IAU	LAC26	13

Feature	Source	Reference	Map(s)
Eudoxus I	IAU	LAC26	13
Eudoxus U	IAU	LAC26	13
Eudoxus V	IAU	LAC26	13
Eudoxus W	A&G	98-2	13
Eudoxus γ	A&G	98-2	13
Euler	IAU	LAC39	19, 20
Euler E	IAU	LAC39	19
Euler F	IAU	LAC40	19, 20
Euler G	IAU	LAC40	20
Euler H	IAU	LAC40	20
Euler J	IAU	LAC39	19
Euler L	IAU	LAC40	19, 20
Euler γ	A&G	133-3	19, 20
Euler δ	A&G	133-3	19, 20
Euler ν	A&G	133-3	19
Euler π	A&G	133-3	19, 20
Explorer	CAF	B6	Detail 6
Exuperay	CAF	B4	Detail 2
Fabbroni	IAU	LAC42	24, 25
Fahrenheit	IAU	LAC62	38
Falcon	IAU	LAC43	Detail 6
Family Mountain	IAU	LAC43	Detail 6
Fan Crater	CAF	B4	Detail 2
Faust	CAF	B6	Detail 6
Fauth	IAU	LAC58	31
Fauth A	IAU	LAC58	31
Fauth B	IAU	LAC58	31
Fauth C	IAU	LAC58	31
Fauth D	IAU	LAC58	31
Fauth E	IAU	LAC58	31
Fauth F	IAU	LAC58	31
Fauth G	IAU	LAC58	31
Fauth H	IAU	LAC58	31
Faye Ridge	CAF	B1	Detail 8
Fedorov	IAU	LAC39	19
Felix	IAU	LAC40	20
Feuillée	IAU	LAC40	21
Fifty Five	CAF	B4	Detail 2
Fifty Four	CAF	B4	Detail 2
Finsch	IAU	LAC42	24
Firmicus	IAU	LAC62	38
Firmicus A	IAU	LAC62	38
Firmicus B	IAU	LAC62	38
Firmicus C	IAU	LAC62	38
Firmicus D	IAU	LAC62	38
Firmicus E	IAU	LAC62	38
Firmicus F	IAU	LAC62	38
Firmicus G	IAU	LAC62	38
Firmicus H	IAU	LAC62	38
Firmicus M	IAU	LAC62	38
Flow Crater	CAF	B4	Detail 2
Fontenelle	IAU	LAC12	3
Fontenelle A	IAU	LAC3	3
Fontenelle B	IAU	LAC11	3
Fontenelle C	IAU	LAC3	3
Fontenelle D	IAU	LAC11	3
Fontenelle E	IAU	LAC3	3
Fontenelle G	IAU	LAC12	3
Fontenelle H	IAU	LAC3	3
Fontenelle K	IAU	LAC3	3, 4
Fontenelle L	IAU	LAC3	3
Fontenelle M	IAU	LAC11	3
Fontenelle N	IAU	LAC3	3
Fontenelle P	IAU	LAC3	3
Fontenelle R	IAU	LAC3	3
Fontenelle S	IAU	LAC3	3
Fontenelle T	IAU	LAC3	3
Fontenelle X	IAU	LAC11	3
Foucault	IAU	LAC11	2
Foucault β	A&G	164-1	2
Foucault γ	A&G	164-1	2
Franck	IAU	LAC43	25
Franklin	IAU	LAC27	15
Franklin C	IAU	LAC27	15
Franklin F	IAU	LAC27	15
Franklin G	IAU	LAC27	15
Franklin H	IAU	LAC27	15
Franklin K	IAU	LAC27	15
Franklin W	IAU	LAC27	15
Franklin γ	A&G	74-2	15
Franz	IAU	LAC43	25
Fredholm	IAU	LAC43	26
Fresnel ψ	A&G	103-1	23
Freud	IAU	LAC38	18
Front Crater	CAF	B4	Detail 2
Frosty	CAF	B6	Detail 6
Furnace Gulch	CAF	B1	Detail 8
G. Bond	IAU	LAC27	15
G. Bond A	IAU	LAC43	15
G. Bond B	IAU	LAC43	15, 25
G. Bond BA	A&G	79-1	15
G. Bond C	IAU	LAC43	25
G. Bond G	IAU	LAC27	15
G. Bond J	A&G	79-1	15
G. Bond K	IAU	LAC27	15
Galen	IAU	LAC41	22
Galilaei	IAU	LAC56	28
Galilaei A	IAU	LAC56	28
Galilaei B	IAU	LAC56	28
Galilaei D	IAU	LAC56	28
Galilaei E	IAU	LAC56	17, 28
Galilaei F	IAU	LAC56	28
Galilaei G	IAU	LAC56	28
Galilaei H	IAU	LAC56	28
Galilaei J	IAU	LAC56	28
Galilaei K	IAU	LAC56	28
Galilaei L	IAU	LAC56	28
Galilaei M	IAU	LAC56	28
Galilaei S	IAU	LAC56	17
Galilaei T	IAU	LAC38	17

Feature	Source	Reference	Map(s)
Galilaei V	IAU	LAC38	17
Galilaei W	IAU	LAC38	17
Galilaei χ	A&G	157-2	28
Galle	IAU	LAC13	5
Galle A	IAU	LAC13	5
Galle B	IAU	LAC13	5
Galle BA	A&G	103-3	5
Galle C	IAU	LAC13	5
Galvani	IAU	LAC21	1
Gambart	IAU	LAC58	31, 32
Gambart A	IAU	LAC58	31
Gambart AA	A&G	121-1	31
Gambart AB	A&G	121-1	31
Gambart AC	A&G	121-1	31
Gambart B	IAU	LAC58	32
Gambart BB	A&G	114-1	32
Gambart BC	A&G	114-1	32
Gambart C	IAU	LAC58	32
Gambart CA	A&G	114-1	32
Gambart CB	A&G	114-1	32
Gambart CC	A&G	114-1	32
Gambart CD	A&G	114-1	32
Gambart D	IAU	LAC58	31
Gambart E	IAU	LAC58	31
Gambart EA	A&G	121-1	31
Gambart F	IAU	LAC58	31
Gambart G	IAU	LAC58	32
Gambart H	IAU	LAC58	32
Gambart J	IAU	LAC76	31
Gambart K	IAU	LAC58	32
Gambart L	IAU	LAC58	31, 32
Gambart M	IAU	LAC58	32
Gambart MA	A&G	114-1	32
Gambart N	IAU	LAC76	32
Gambart NA	A&G	121-1	31
Gambart R	IAU	LAC76	31
Gambart S	IAU	LAC76	32
Gambart θ	A&G	121-1	31
Gambart λ	A&G	121-1	31
Gambart ν	A&G	121-1	31
Gambart ρ	A&G	121-1	31
Gambart τ	A&G	121-1	31
Gambart ω	A&G	121-1	31
Gardner	IAU	LAC43	25
Gärtner	IAU	LAC13	5, 6
Gärtner A	IAU	LAC13	6
Gärtner C	IAU	LAC13	5
Gärtner D	IAU	LAC13	6
Gärtner E	IAU	LAC14	6
Gärtner F	IAU	LAC13	5, 6
Gärtner FA	A&G	92-1	5, 6
Gärtner G	IAU	LAC13	6
Gärtner M	IAU	LAC13	6
Gaston	IAU	LAC39	9
Gateway	CAF	B4	Detail 2

Feature	Source	Reference	Map(s)
Gatsby	CAF	B6	Detail 6
Gauss	IAU	LAC28	16
Gauss A	IAU	LAC29	16
Gauss B	IAU	LAC28	16
Gauss C	IAU	LAC28	16
Gauss D	IAU	LAC28	16
Gauss E	IAU	LAC28	16
Gauss F	IAU	LAC28	16
Gauss G	IAU	LAC28	16
Gauss H	IAU	LAC28	16
Gauss J	IAU	LAC28	16
Gauss W	IAU	LAC28	16
Gay-Lussac	IAU	LAC58	20, 31
Gay-Lussac A	IAU	LAC58	31
Gay-Lussac B	IAU	LAC40	20
Gay-Lussac C	IAU	LAC58	20
Gay-Lussac D	IAU	LAC58	20, 31
Gay-Lussac F	IAU	LAC58	20, 31
Gay-Lussac G	IAU	LAC58	20, 31
Gay-Lussac H	IAU	LAC58	31
Gay-Lussac J	IAU	LAC58	31
Gay-Lussac M	A&G	126-2	31
Gay-Lussac N	IAU	LAC58	31
Gemini Ridge	CAF	B1	Detail 8
Geminus	IAU	LAC27	16
Geminus A	IAU	LAC44	16
Geminus B	IAU	LAC27	16
Geminus C	IAU	LAC28	16
Geminus D	IAU	LAC43	15, 16
Geminus DA	A&G	67-1	15
Geminus E	IAU	LAC27	15
Geminus EA	A&G	67-1	15
Geminus EB	A&G	67-1	15
Geminus F	IAU	LAC27	16
Geminus G	IAU	LAC43	15, 16
Geminus H	IAU	LAC43	15, 16
Geminus M	IAU	LAC43	15, 16
Geminus N	IAU	LAC43	15, 16
Geminus W	IAU	LAC27	15
Geminus Z	IAU	LAC43	15
Gena	IAU	LAC24	Detail 1
Gerard	IAU	LAC22	8
Gerard A	IAU	LAC22	8
Gerard C	IAU	LAC22	8
Gerard D	IAU	LAC22	8
Gerard E	IAU	LAC22	8
Gerard F	IAU	LAC22	8
Gerard K	IAU	LAC22	8
Gerard L	IAU	LAC22	8
Gerard Q Inner	IAU	LAC22	8
Gerard Q Outer	IAU	LAC22	8
Ghostbead	CAF	B4	Detail 2
Gilbert P	IAU	LAC81	38
Gilbert V	IAU	LAC81	38
Gilbert W	IAU	LAC81	38

Feature	Source	Reference	Map(s)
Gioia	IAU	LAC1	4
Glaisher	IAU	LAC61	37
Glaisher A	IAU	LAC62	37
Glaisher B	IAU	LAC62	37
Glaisher E	IAU	LAC61	37
Glaisher F	IAU	LAC61	26, 37
Glaisher G	IAU	LAC61	37
Glaisher H	IAU	LAC61	26, 37
Glaisher L	IAU	LAC61	37
Glaisher M	IAU	LAC61	37
Glaisher N	IAU	LAC61	37
Glaisher V	IAU	LAC62	37
Glaisher W	IAU	LAC61	37
Glaisher X	A&G	61-2	26
Glushko	IAU	LAC55	28
Goddard B	IAU	LAC45	27
Goddard C	IAU	LAC45	27
Godin	IAU	LAC60	34
Godin A	IAU	LAC59	34
Godin B	IAU	LAC59	34
Godin C	IAU	LAC59	34
Godin D	IAU	LAC59	34
Godin E	IAU	LAC60	34
Godin G	IAU	LAC60	34
Godin β	AIC	60D	34
Goldschmidt	IAU	LAC3	4
Goldschmidt A	IAU	LAC3	4
Goldschmidt B	IAU	LAC3	4
Goldschmidt C	IAU	LAC3	4
Goldschmidt D	IAU	LAC3	4
Golgi	IAU	LAC38	18
Grace	IAU	LAC61	25, 36
Greaves	IAU	LAC62	37
Grove	IAU	LAC26	14
Grove Y	IAU	LAC26	14
Gruithuisen	IAU	LAC23	9
Gruithuisen B	IAU	LAC23	9
Gruithuisen E	IAU	LAC23	9
Gruithuisen F	IAU	LAC24	9
Gruithuisen G	IAU	LAC23	9
Gruithuisen H	IAU	LAC23	9
Gruithuisen K	IAU	LAC23	9
Gruithuisen M	IAU	LAC23	9
Gruithuisen P	IAU	LAC23	9
Gruithuisen R	IAU	LAC23	9
Gruithuisen S	IAU	LAC23	9
Gruithuisen ζ	A&G	151-2	9
Guang Han Gong	IAU	LAC24	11
Hadley C	IAU	LAC41	22
Hadley δ	A&G	102-3	22
Hahn	IAU	LAC45	16
Hahn A	IAU	LAC44	16, 26, 27
Hahn AD	NLC	LAC44	26
Hahn B	IAU	LAC45	16
Hahn D	IAU	LAC44	26, 27
Hahn DA	NLC	LAC44	26, 27
Hahn DB	NLC	LAC44	26, 27
Hahn E	IAU	LAC44	27
Hahn F	IAU	LAC28	16
Hall	IAU	LAC27	15
Hall C	IAU	LAC27	15
Hall J	IAU	LAC27	15
Hall K	IAU	LAC27	14, 15
Hall X	IAU	LAC27	15
Hall Y	IAU	LAC27	15
Hanover	CAF	B6	Detail 6
Hansen	IAU	LAC63	27, 38
Hansen A	IAU	LAC63	38
Hansen B	IAU	LAC63	27, 38
Harbringer α	A&G	144-3	19
Harbringer β	A&G	144-3	19
Harbringer γ	A&G	144-3	19
Harbringer δ	A&G	144-3	19
Harbringer ζ	A&G	144-3	19
Harbringer θ	A&G	144-3	19
Harbringer μ	A&G	144-3	19
Harding	IAU	LAC22	8
Harding A	IAU	LAC22	8
Harding B	IAU	LAC22	8
Harding C	IAU	LAC22	8
Harding D	IAU	LAC22	8
Harding H	IAU	LAC22	8
Harpalus	IAU	LAC11	2
Harpalus B	IAU	LAC11	2
Harpalus C	IAU	LAC11	2
Harpalus E	IAU	LAC10	1
Harpalus G	IAU	LAC10	1
Harpalus H	IAU	LAC10	1
Harpalus S	IAU	LAC10	1
Harpalus T	IAU	LAC11	1
Harpalus γ	A&G	164-1	1, 2
Harpalus λ	A&G	164-1	2
Harpalus ξ	A&G	164-1	2
Hayn	IAU	LAC5	6
Hayn A	IAU	LAC15	6
Hayn B	IAU	LAC5	6
Hayn D	IAU	LAC5	6
Hayn E	IAU	LAC5	6
Hayn F	IAU	LAC5	6
Hayn H	IAU	LAC14	6
Hayn J	IAU	LAC5	6
Hayn L	IAU	LAC5	6
Hayn M	IAU	LAC14	6
Hayn S	IAU	LAC5	6
Hedin	IAU	LAC55	28
Hedin A	IAU	LAC55	28
Hedin B	IAU	LAC55	28
Hedin C	IAU	LAC55	28
Hedin F	IAU	LAC55	28
Hedin G	IAU	LAC55	28

Feature	Source	Reference	Map(s)
Hedin H	IAU	LAC55	28
Hedin K	IAU	LAC55	28
Hedin L	IAU	LAC55	28
Hedin N	IAU	LAC55	28
Hedin R	IAU	LAC55	28
Hedin S	IAU	LAC55	28
Hedin T	IAU	LAC55	28
Hedin V	IAU	LAC55	28
Hedin Z	IAU	LAC55	28
Heinrich	IAU	LAC40	21
Heis	IAU	LAC24	9, 10
Heis A	IAU	LAC24	9, 10
Heis D	IAU	LAC39	9, 10
Helicon	IAU	LAC24	10
Helicon B	IAU	LAC24	10
Helicon BA	A&G	134-2	10
Helicon C	IAU	LAC24	10
Helicon C*	IAU	LAC24	10
Helicon E	IAU	LAC24	10
Helicon G	IAU	LAC24	10
Henry	IAU	B6	Detail 6
Henson	IAU	B6	Detail 6
Heraclides A	IAU	LAC24	10
Heraclides E	IAU	LAC24	10
Heraclides F	IAU	LAC24	10
Hercules	IAU	LAC27	6, 14
Hercules B	IAU	LAC27	6, 14
Hercules C	IAU	LAC27	14
Hercules D	IAU	LAC27	14, 15
Hercules E	IAU	LAC27	14
Hercules F	IAU	LAC14	6
Hercules G	IAU	LAC27	14
Hercules H	IAU	LAC14	6
Hercules J	IAU	LAC27	14
Hercules K	IAU	LAC27	14
Hermann	IAU	LAC74	28
Hermann A	IAU	LAC56	28
Hermann B	IAU	LAC74	28
Hermann BA	A&G	169-1	28
Hermann C	IAU	LAC74	28
Hermann E	IAU	LAC56	29
Hermann F	IAU	LAC56	28
Hermann H	IAU	LAC56	28
Hermann J	IAU	LAC56	28
Hermann K	IAU	LAC56	28
Hermann L	IAU	LAC56	28
Hermann R	IAU	LAC56	28
Hermann S	IAU	LAC56	28
Herodotus	IAU	LAC39	18
Herodotus A	IAU	LAC38	18
Herodotus B	IAU	LAC38	18
Herodotus C	IAU	LAC38	18
Herodotus E	IAU	LAC38	8, 18
Herodotus G	IAU	LAC38	18
Herodotus H	IAU	LAC38	18
Herodotus K	IAU	LAC38	18
Herodotus L	IAU	LAC38	18
Herodotus N	IAU	LAC38	18
Herodotus R	IAU	LAC38	18
Herodotus S	IAU	LAC38	18
Herodotus T	IAU	LAC38	18
Herodotus γ	A&G	158-1	18
Herodotus δ	A&G	150-3	18
Herodotus η	A&G	158 1	18
Herodotus θ	A&G	157-3	18
Herodotus ι	A&G	158-1	18
Herodotus κ	A&G	158-1	18
Herodotus λ	A&G	158-1	18
Herodotus ν	A&G	158-1	18
Herodotus ρ	A&G	150-3	18
Herodotus σ	A&G	158-1	18
Herodotus τ	A&G	150-3	18
Herodotus ω	A&G	150-3	18
Hess-Apollo	IAU	LAC43	Detail 6
Hevelius	IAU	LAC56	28
Hevelius A	IAU	LAC56	28
Hevelius B	IAU	LAC56	28
Hevelius D	IAU	LAC56	28
Hevelius E	IAU	LAC56	28
Hevelius F	A&G	169-1	28
Hevelius G	A&G	169-1	28
Hevelius H	A&G	169-1	28
Hevelius J	IAU	LAC56	28
Hevelius K	IAU	LAC55	28
Hevelius L	IAU	LAC55	28
Hevelius α	A&G	169-1	28
High	CAF	B4	Detail 2
Hill	IAU	LAC43	25
Holden	IAU	B6	Detail 6
Hole-in-the-Wall	CAF	B6	Detail 6
Hooke	IAU	LAC27	15
Hooke D	IAU	LAC27	15
Horatio	IAU	LAC43	Detail 6
Hornsby	IAU	LAC42	23
Horrebow	IAU	LAC11	2
Horrebow A	IAU	LAC11	2
Horrebow B	IAU	LAC11	2
Horrebow C	IAU	LAC11	2
Horrebow D	IAU	LAC11	2
Horrebow G	IAU	LAC11	2
Hortensius	IAU	LAC58	30
Hortensius A	IAU	LAC57	30
Hortensius B	IAU	LAC58	30
Hortensius BB	AIC	58D	30
Hortensius C	IAU	LAC58	30, 31
Hortensius D	IAU	LAC57	30
Hortensius DA	A&G	133-1	30
Hortensius DC	A&G	133-1	30
Hortensius DD	A&G	133-1	30
Hortensius E	IAU	LAC58	31

Feature	Source	Reference	Map(s)
Hortensius EA	A&G	133-1	30, 31
Hortensius EB	A&G	133-1	30, 31
Hortensius EC	A&G	133-1	30, 31
Hortensius F	IAU	LAC58	31
Hortensius G	IAU	LAC58	30, 31
Hortensius H	IAU	LAC57	30
Hortensius β	A&G	133-1	30, 31
Hortensius γ	A&G	133-1	30, 31
Hortensius δ	A&G	126-1	31
Hortensius ϵ	A&G	126-2	31
Hortensius ρ	A&G	133-1	30
Hortensius σ	A&G	133-1	30
Hortensius τ	A&G	133-1	30
Hortensius ϕ	A&G	133-1	30, 31
Hortensius ω	A&G	133-1	30
Hubble	IAU	LAC45	27
Humason	IAU	LAC38	8
Huxley	IAU	LAC41	22
Huygens A	IAU	LAC41	22
Huygens M	A&G	109-2	22
Huygens β	A&G	109-2	22
Hyginus	IAU	LAC59	34
Hyginus A	IAU	LAC59	33, 34
Hyginus B	IAU	LAC59	33, 34
Hyginus C	IAU	LAC59	34
Hyginus D	IAU	LAC59	33
Hyginus E	IAU	LAC59	34
Hyginus F	IAU	LAC59	34
Hyginus G	IAU	LAC59	34
Hyginus H	IAU	LAC59	34
Hyginus N	IAU	LAC59	34
Hyginus NA	A&G	97-2	34
Hyginus S	IAU	LAC59	34
Hyginus W	IAU	LAC59	34
Hyginus Z	IAU	LAC59	34
Hypatia C	IAU	LAC78	35
Hypatia CA	A&G	85-1	35
Hypatia CC	AIC	78B	35
Hypatia CD	AIC	78B	35
Hypatia E	IAU	LAC78	35
Ian	IAU	LAC41	22
Icarus	CAF	B4	Detail 2
Igor	IAU	LAC24	Detail 1
Ina	IAU	LAC41	Detail 3
Index	IAU	LAC41	Detail 2
Isabel	IAU	LAC39	19
Isis	IAU	LAC42	Detail 5
Ivan	IAU	LAC39	19
J. Herschel	IAU	LAC11	2
J. Herschel B	IAU	LAC11	2
J. Herschel C	IAU	LAC11	2
J. Herschel D	IAU	LAC11	2
J. Herschel F	IAU	LAC11	2
J. Herschel G	A&G	164-1	2
J. Herschel K	IAU	LAC11	2

Feature	Source	Reference	Map(s)
J. Herschel L	IAU	LAC11	2
J. Herschel M	IAU	LAC11	[illegible]
J. Herschel N	IAU	LAC11	2, 3
J. Herschel P	IAU	LAC11	3
J. Herschel R	IAU	LAC11	3
Jansen	IAU	LAC60	25, 35, 36
Jansen D	IAU	LAC60	24, 25
Jansen E	IAU	LAC60	24, 25, 35, 36
Jansen EA	A&G	85-2	24, 35
Jansen EB	A&G	78-2	24, 24
Jansen G	IAU	LAC60	35
Jansen H	IAU	LAC60	35, 36
Jansen K	IAU	LAC60	36
Jansen L	IAU	LAC61	25
Jansen R	IAU	LAC60	24, 25
Jansen T	IAU	LAC61	36
Jansen U	IAU	LAC61	36
Jansen W	IAU	LAC60	36
Jansen Y	IAU	LAC60	35, 36
Jansen δ	A&G	78-2	36
Jehan	IAU	LAC39	19
Jenkins	IAU	LAC63	38
Jerik	IAU	LAC42	Detail 5
Jomo	IAU	LAC41	22
Jones	CAF	B6	Detail 6
Joy	IAU	LAC41	23
Julienne	IAU	LAC41	22
Julius Caesar	IAU	LAC60	34
Julius Caesar A	IAU	LAC60	34
Julius Caesar AB	AIC	60D	34
Julius Caesar AC	AIC	60D	34
Julius Caesar B	IAU	LAC60	34
Julius Caesar C	IAU	LAC60	34
Julius Caesar D	IAU	LAC60	34, 35
Julius Caesar F	IAU	LAC60	34
Julius Caesar G	IAU	LAC60	34
Julius Caesar H	IAU	LAC60	34
Julius Caesar J	IAU	LAC60	34
Julius Caesar P	IAU	LAC60	34
Julius Caesar PA	A&G	90-2	34
Julius Caesar Q	IAU	LAC60	34
Julius Caesar η	AIC	60D	34, 35
Kane	IAU	LAC13	5
Kane A	IAU	LAC13	5
Kane F	IAU	LAC13	5
Kane G	IAU	LAC13	5
Kathleen	IAU	LAC41	22
Keldysh	IAU	LAC14	6
Kepler	IAU	LAC57	30
Kepler A	IAU	LAC57	30
Kepler B	IAU	LAC57	30
Kepler C	IAU	LAC57	29
Kepler CA	A&G	144-2	29
Kepler CB	A&G	144-2	29, 30
Kepler D	IAU	LAC57	29

Feature	Source	Reference	Map(s)
Kepler E	IAU	LAC57	29
Kepler F	IAU	LAC57	30
Kepler P	IAU	LAC57	30
Kepler T	IAU	LAC57	30
Kepler δ	A&G	138-2	30
Kepler ϵ	A&G	138-2	30
Kepler η	A&G	138-2	30
Kepler θ	A&G	133-1	30
Kepler κ	A&G	144-1	29, 30
Kepler λ	A&G	138-2	30
Kepler μ	A&G	138-2	30
Kepler ν	A&G	138-2	30
Kepler π	A&G	144-2	29
Kepler ϕ	A&G	138-2	30
Kimbal Crater	CAF	B4	Detail 2
Kirch	IAU	LAC25	12
Kirch E	IAU	LAC25	11, 12
Kirch F	IAU	LAC25	11, 12
Kirch G	IAU	LAC25	11
Kirch H	IAU	LAC25	11, 12
Kirch K	IAU	LAC25	12
Kirch M	IAU	LAC25	11
Kirchhoff	IAU	LAC43	15
Kirchhoff C	IAU	LAC43	15
Kirchhoff E	IAU	LAC43	15
Kirchhoff F	IAU	LAC43	15
Kirchhoff G	IAU	LAC43	15, 25
Knox-Shaw	IAU	LAC63	38
Kolya	IAU	LAC24	Detail 1
Kostya	IAU	LAC24	Detail 1
Krafft	IAU	LAC37	17
Krafft C	IAU	LAC37	17
Krafft D	IAU	LAC55	17
Krafft E	IAU	LAC37	17
Krafft H	IAU	LAC37	17
Krafft K	IAU	LAC37	17
Krafft L	IAU	LAC37	17
Krafft M	IAU	LAC37	17
Krafft U	IAU	LAC38	17
Krieger	IAU	LAC38	18, 19
Krieger C	IAU	LAC39	19
Krogh	IAU	LAC62	38
Kunowsky	IAU	LAC57	30
Kunowsky C	IAU	LAC75	30
Kunowsky CA	A&G	132-3	30
Kunowsky D	IAU	LAC58	30
Kunowsky G	IAU	LAC57	30
Kunowsky H	IAU	LAC58	30
Kunowsky κ	A&G	133-1	30
Kunowsky σ	A&G	133-1	30
Kunowsky ω	A&G	133-1	30
La Condamine	IAU	LAC11	2
La Condamine A	IAU	LAC11	2
La Condamine B	IAU	LAC11	2, 3
La Condamine C	IAU	LAC11	2
La Condamine D	IAU	LAC11	2
La Condamine E	IAU	LAC11	2
La Condamine F	IAU	LAC11	2
La Condamine G	IAU	LAC11	2, 3
La Condamine H	IAU	LAC11	2, 3
La Condamine J	IAU	LAC12	3
La Condamine JA	A&G	128-1	3
La Condamine K	IAU	LAC11	2, 3
La Condamine L	IAU	LAC11	2, 3
La Condamine M	IAU	LAC11	2, 3
La Condamine N	IAU	LAC11	3
La Condamine O	IAU	LAC11	3
La Condamine P	IAU	LAC11	3
La Condamine Q	IAU	LAC11	3
La Condamine R	IAU	LAC11	3
La Condamine S	IAU	LAC11	3
La Condamine SA	A&G	140-1	3
La Condamine T	IAU	LAC11	3
La Condamine TA	A&G	140-1	3
La Condamine U	IAU	LAC11	3
La Condamine V	IAU	LAC11	3
La Condamine X	IAU	LAC11	3
La Hire A	IAU	LAC40	20
La Hire B	IAU	LAC40	20
Lacus Bonitatis	IAU	LAC43	25, 26
Lacus Doloris	IAU	LAC41	23
Lacus Felicitatis	IAU	LAC41	22, 23
Lacus Gaudii	IAU	LAC42	23
Lacus Hiemalis	IAU	LAC60	23
Lacus Lenitatis	IAU	LAC60	23, 34
Lacus Mortis	IAU	LAC26	14
Lacus Odii	IAU	LAC41	23
Lacus Perseverantiae	IAU	LAC62	38
Lacus Somniorum	IAU	LAC26	14, 15
Lacus Spei	IAU	LAC28	16
Lacus Temporis	IAU	LAC27	7, 15
Lade	IAU	LAC77	34
Lade A	IAU	LAC60	34
Lade B	IAU	LAC59	34
Lade C	A&G	97-1	34
Lade D	IAU	LAC78	34
Lade M	IAU	LAC77	34
Lade S	IAU	LAC77	34
Lade T	IAU	LAC77	34
Lade U	IAU	LAC77	34
Lade V	IAU	LAC77	34
Lade W	IAU	LAC59	34
Lade λ	A&G	97-1	34
Lambert	IAU	LAC40	20
Lambert A	IAU	LAC40	20
Lambert B	IAU	LAC40	20
Lambert R	IAU	LAC40	20
Lambert T	IAU	LAC40	20
Lambert W	IAU	LAC40	20
Lambert γ	A&G	126-3	20

Feature	Source	Reference	Map(s)
Lambert δ	A&G	121-3	20, 21
[illegible]	[illegible]	[illegible]	[illegible]
Lamont	IAU	LAC60	35
Landsteiner	IAU	LAC40	11
Langley J	IAU	LAC21	1
Lansberg	IAU	LAC58	30, 31
Lansberg A	IAU	LAC57	30
Lansberg AA	A&G	132-3	30
Lansberg AB	A&G	132-3	30
Lansberg G	IAU	LAC76	30
Lansberg GA	A&G	132-3	30
Lansberg X	IAU	LAC58	30
Lansberg Y	IAU	LAC58	30
Lansberg α	A&G	126-1	31
Lansberg β	A&G	126-1	31
Lansberg λ	A&G	126-1	31
Lansberg π	A&G	126-1	31
Lansberg ϕ	A&G	132-3	30
Laplace A	IAU	LAC24	10
Laplace B	IAU	LAC12	3
Laplace D	IAU	LAC24	10
Laplace E	IAU	LAC11	3
Laplace F	IAU	LAC24	11
Laplace FA	A&G	127-2	11
Laplace HA	A&G	140-1	3
Laplace L	IAU	LAC11	3
Laplace M	IAU	LAC12	3
Lara	IAU	LAC43	Detail 6
Last	IAU	LAC41	Detail 2
Last Ridge	CAF	B1	Detail 8
Lavoisier	IAU	LAC22	8
Lavoisier A	IAU	LAC22	8
Lavoisier B	IAU	LAC22	8
Lavoisier C	IAU	LAC22	8
Lavoisier E	IAU	LAC22	8
Lavoisier F	IAU	LAC22	8
Lavoisier G	IAU	LAC22	8
Lavoisier H	IAU	LAC22	8
Lavoisier J	IAU	LAC36	8
Lavoisier K	IAU	LAC22	8
Lavoisier L	IAU	LAC22	8
Lavoisier N	IAU	LAC22	8
Lavoisier S	IAU	LAC22	8
Lavoisier T	IAU	LAC22	8
Lavoisier W	IAU	LAC22	8
Lavoisier Z	IAU	LAC36	8
Lawrence	IAU	LAC61	37
Le Monnier	IAU	LAC42	24, 25
Le Monnier A	IAU	LAC43	25
Le Monnier BL	A&G	86-1	24
Le Monnier BM	A&G	86-1	24
Le Monnier H	IAU	LAC42	24
Le Monnier K	IAU	LAC43	24
Le Monnier KA	A&G	86-1	24
Le Monnier KB	A&G	86-1	24, 25
Le Monnier L	NLC	LAC42	24
[illegible]	[illegible]	[illegible]	[illegible]
Le Monnier LB	A&G	78-3	24
Le Monnier LC	A&G	86-1	24
Le Monnier S	IAU	LAC43	25
Le Monnier T	IAU	LAC43	25
Le Monnier U	IAU	LAC43	25
Le Monnier V	IAU	LAC43	25
Le Monnier α	NLC	LAC42	24
Le Verrier	IAU	LAC24	10, 11
Le Verrier A	IAU	LAC24	11
Le Verrier B	IAU	LAC25	11
Le Verrier D	IAU	LAC25	11
Le Verrier E	IAU	LAC24	11
Le Verrier S	IAU	LAC24	10, 11
Le Verrier T	IAU	LAC24	10, 11
Le Verrier U	IAU	LAC25	11
Le Verrier V	IAU	LAC24	11
Le Verrier W	IAU	LAC25	11
Le Verrier X	IAU	LAC25	11
Lee Lincoln Scarp	CAF	B6	Detail 6
Leonid	IAU	LAC24	Detail 1
Lev	IAU	LAC62	38
Lewis	IAU	B6	Detail 6
Lichtenberg	IAU	LAC22	8
Lichtenberg A	IAU	LAC38	18
Lichtenberg AA	A&G	163-1	18
Lichtenberg B	IAU	LAC23	8
Lichtenberg F	IAU	LAC22	8
Lichtenberg H	IAU	LAC38	8
Lichtenberg R	IAU	LAC22	8
Lichtenberg β	A&G	175-1	8
Lichtenberg ϵ	A&G	175-1	8
Lick	IAU	LAC62	37
Lick A	IAU	LAC62	37
Lick B	IAU	LAC62	37
Lick BA	A&G	61-2	37
Lick C	IAU	LAC62	37
Lick E	IAU	LAC62	37
Lick F	IAU	LAC62	37
Lick G	IAU	LAC62	37
Lick K	IAU	LAC62	37
Lick L	IAU	LAC61	37
Lick N	IAU	LAC61	37
Light Mantle	IAU	LAC43	Detail 6
Lightning	CAF	B4	Detail 2
Linda	IAU	LAC39	9
Link	CAF	B4	Detail 2
Linné	IAU	LAC42	23
Linné A	IAU	LAC42	23
Linné AB	A&G	98-1	23
Linné AC	A&G	98-1	23
Linné AD	A&G	98-1	23
Linné AE	A&G	97-3	23
Linné B	IAU	LAC42	13

Feature	Source	Reference	Map(s)
Linné BA	A&G	103-1	13
Linné BB	A&G	103-1	13
Linné BC	A&G	98-1	13
Linné D	IAU	LAC42	23
Linné EA	A&G	90-3	23, 24
Linné EB	A&G	90-3	23, 24
Linné EC	A&G	90-3	23, 24
Linné ED	A&G	98-1	23, 23, 24
Linné EF	A&G	90-3	23
Linné F	IAU	LAC26	13
Linné G	IAU	LAC26	13
Linné H	IAU	LAC26	13
Liouville	IAU	LAC63	38
Little Moltke	CAF	B1	Detail 8
Little West	IAU	LAC60	Detail 7
Littrow	IAU	LAC43	25
Littrow A	IAU	LAC43	25
Littrow BA	A&G	78-3	24, 25
Littrow BB	A&G	78-3	24
Littrow BC	A&G	78-3	25
Littrow D	IAU	LAC43	25
Littrow F	IAU	LAC43	25
Littrow P	IAU	LAC43	25
Liu Hui	IAU	LAC23	9
Locke	CAF	B6	Detail 6
Lohrmann	IAU	LAC74	28
Lohrmann A	IAU	LAC74	28
Lohrmann B	IAU	LAC74	28
Lohrmann D	IAU	LAC74	28
Lohrmann M	IAU	LAC74	28
Lohrmann N	IAU	LAC73	28
Lonely	CAF	B4	Detail 2
Lonesome Mesa	CAF	B1	Detail 8
Lost Basin	CAF	B1	Detail 8
Lost Crater	CAF	B1	Detail 8
Lost Valley	CAF	B1	Detail 8
Louise	IAU	LAC39	19
Louville	IAU	LAC23	9
Louville A	IAU	LAC23	9
Louville B	IAU	LAC23	9, 9
Louville D	IAU	LAC23	9
Louville DA	IAU	LAC23	9
Louville E	IAU	LAC23	9
Louville K	IAU	LAC23	9
Louville P	IAU	LAC23	9
Low Mesa	CAF	B1	Detail 8
Lubbock M	IAU	LAC79	36
Lubbock R	IAU	LAC79	37
Lubbock S	IAU	LAC61	37
Lucian	IAU	LAC61	25, 36
Luke	CAF	B4	Detail 2
Lundi	CAF	B4	Detail 2
Luther	IAU	LAC26	14
Luther H	IAU	LAC26	14
Luther K	IAU	LAC26	14
Luther X	IAU	LAC26	14
Luther Y	IAU	LAC26	14
Luther α	A&G	86-1	14
Lyell	IAU	LAC61	25, 36
Lyell A	IAU	LAC61	25, 36
Lyell B	IAU	LAC61	25, 36
Lyell C	IAU	LAC61	25
Lyell D	IAU	LAC61	25, 26
Lyell K	IAU	LAC61	25
Mackin	IAU	LAC43	Detail 6
Maclaurin H	IAU	LAC80	38
Maclaurin K	IAU	LAC80	38
Maclaurin L	IAU	LAC80	38
Maclaurin O	IAU	LAC80	38
Maclaurin W	IAU	LAC62	38
Maclaurin X	IAU	LAC62	38
Maclear	IAU	LAC60	35
Maclear A	IAU	LAC60	35
Macmillan	IAU	LAC41	21
Macrobius	IAU	LAC43	26
Macrobius AA	A&G	66-3	25
Macrobius AB	A&G	73-3	25
Macrobius BA	A&G	73-3	25
Macrobius C	IAU	LAC43	26
Macrobius E	IAU	LAC43	26
Macrobius F	IAU	LAC43	26
Macrobius K	IAU	LAC43	25
Macrobius M	IAU	LAC43	25
Macrobius N	IAU	LAC43	25
Macrobius P	IAU	LAC43	25
Macrobius Q	IAU	LAC43	26
Macrobius S	IAU	LAC43	26
Macrobius T	IAU	LAC43	26
Macrobius U	IAU	LAC43	25
Macrobius V	IAU	LAC43	25
Macrobius W	IAU	LAC43	25, 26
Macrobius X	IAU	LAC43	25
Macrobius Y	IAU	LAC43	25
Macrobius Z	IAU	LAC43	25
Macrobius ZA	A&G	66-3	25
Macrobius ZB	A&G	66-3	25
Maestlin	IAU	LAC57	29
Maestlin G	IAU	LAC57	29
Maestlin H	IAU	LAC57	29
Maestlin R	IAU	LAC57	29
Maestlin ι	A&G	138-1	29
Maestlin λ	A&G	144-1	29
Maestlin μ	A&G	144-1	29
Maestlin ν	A&G	144-1	29
Main	IAU	LAC1	4
Main L	IAU	LAC1	4
Main N	IAU	LAC1	4
Mairan	IAU	LAC23	9
Mairan A	IAU	LAC23	9
Mairan C	IAU	LAC23	9

Feature	Source	Reference	Map(s)
Mairan D	IAU	LAC23	9
Mairan E	IAU	LAC23	9
Mairan F	IAU	LAC23	9
Mairan G	IAU	LAC23	9
Mairan H	IAU	LAC23	9
Mairan K	IAU	LAC23	9
Mairan L	IAU	LAC23	9
Mairan N	IAU	LAC23	9
Mairan T	IAU	LAC23	9
Mairan Y	IAU	LAC23	9
Manilius	IAU	LAC59	23, 34
Manilius B	IAU	LAC41	23
Manilius C	IAU	LAC60	34
Manilius D	IAU	LAC59	34
Manilius DA	A&G	97-2	34
Manilius DB	A&G	97-2	34
Manilius E	IAU	LAC41	23
Manilius EA	A&G	97-3	23
Manilius FA	A&G	97-2	23
Manilius G	IAU	LAC59	23
Manilius GA	A&G	97-2	23
Manilius H	IAU	LAC41	23
Manilius K	IAU	LAC60	34
Manilius N	A&G	97-2	23, 34
Manilius T	IAU	LAC60	34
Manilius U	IAU	LAC60	23, 34
Manilius W	IAU	LAC60	34
Manilius X	IAU	LAC60	23, 34
Manilius Z	IAU	LAC42	23
Manilius α	A&G	97-2	23
Manilius β	A&G	102-2	22, 23
Manners	IAU	LAC60	35
Manners A	IAU	LAC60	35
Manuel	IAU	LAC42	Detail 4
Maraldi	IAU	LAC43	25
Maraldi A	IAU	LAC43	25
Maraldi D	IAU	LAC43	25
Maraldi E	IAU	LAC43	25
Maraldi F	IAU	LAC43	25
Maraldi N	IAU	LAC43	25
Maraldi R	IAU	LAC43	25
Maraldi W	IAU	LAC61	36
Marco Polo	IAU	LAC59	22
Marco Polo A	IAU	LAC59	22
Marco Polo B	IAU	LAC41	22
Marco Polo C	IAU	LAC59	21, 22, 32, 33
Marco Polo D	IAU	LAC59	22
Marco Polo E	A&G	109-2	21
Marco Polo F	IAU	LAC59	22
Marco Polo G	IAU	LAC41	22
Marco Polo H	IAU	LAC41	22
Marco Polo J	IAU	LAC41	22
Marco Polo K	IAU	LAC41	22
Marco Polo L	IAU	LAC59	21, 22
Marco Polo M	IAU	LAC41	22
Marco Polo P	IAU	LAC41	22
Marco Polo S	IAU	LAC41	22
Marco Polo T	IAU	LAC59	22, 33
Marco Polo γ	A&G	109-2	22
Mare Anguis	IAU	LAC44	27
Mare Crisium	IAU	LAC44	26, 37
Mare Crisium ρ	NLC	LAC44	26
Mare Fecunditatis	IAU	LAC80	37, 38
Mare Frigoris	IAU	LAC11	2, 4, 5, 6
Mare Humboldtianum	IAU	LAC15	7
Mare Imbrium	IAU	LAC24	10, 11, 12, 20, 21
Mare Insularum	IAU	LAC57	30, 31
Mare Marginis	IAU	LAC63	27, 38
Mare Serenitatis	IAU	LAC26	13, 14, 23
Mare Smythii	IAU	LAC63	38
Mare Spumans	IAU	LAC62	38
Mare Tranquillitatis	IAU	LAC61	35, 36
Mare Undarum	IAU	LAC62	38
Mare Vaporum	IAU	LAC41	22, 23, 33, 34
Mariner	CAF	B6	Detail 6
Marius	IAU	LAC56	29
Marius A	IAU	LAC57	29
Marius B	IAU	LAC39	18
Marius BA	A&G	150-2	18
Marius BB	A&G	150-2	18
Marius BC	A&G	150-2	18
Marius C	IAU	LAC57	18, 29
Marius CA	A&G	150-2	29
Marius D	IAU	LAC57	29
Marius DA	A&G	144-2	29
Marius DB	A&G	144-2	29
Marius E	IAU	LAC56	29
Marius F	IAU	LAC57	29
Marius G	IAU	LAC56	29
Marius H	IAU	LAC56	29
Marius J	IAU	LAC57	29
Marius K	IAU	LAC56	29
Marius L	IAU	LAC56	18
Marius M	IAU	LAC38	18
Marius N	IAU	LAC38	18
Marius P	IAU	LAC38	18
Marius Q	IAU	LAC38	18
Marius R	IAU	LAC56	29
Marius S	IAU	LAC57	18, 29
Marius U	IAU	LAC57	29
Marius V	IAU	LAC57	29
Marius W	IAU	LAC57	29
Marius X	IAU	LAC56	28, 29
Marius Y	IAU	LAC56	29
Marius β	A&G	150-2	29
Marius γ	A&G	157-2	18, 29
Marius δ	A&G	150-2	18
Marius ζ	A&G	157-2	18
Marius η	A&G	150-2	29
Marius θ	A&G	157-2	18

Feature	Source	Reference	Map(s)
Marius ι	A&G	157-2	28
Marius κ	A&G	150-2	29
Marius μ	A&G	150-2	18, 29
Marius ν	A&G	150-2	29
Marius ξ	A&G	150-2	29
Marius ρ	A&G	150-2	29
Marius σ	A&G	157-2	28
Marius φ	A&G	157-2	18
Mark	CAF	B4	Detail 2
Markov	IAU	LAC10	1
Markov E	IAU	LAC10	1
Markov F	IAU	LAC10	1
Markov G	IAU	LAC10	1
Markov U	IAU	LAC10	1
Markov θ	A&G	176-1	1
Markov λ	A&G	176-1	1
Markov μ	A&G	176-1	1
Markov σ	A&G	170-3	1
Markov τ	A&G	176-1	1
Mary	IAU	LAC42	Detail 5
Maskelyne	IAU	LAC61	36
Maskelyne A	IAU	LAC61	36
Maskelyne B	IAU	LAC60	36
Maskelyne C	IAU	LAC61	36
Maskelyne D	IAU	LAC61	36
Maskelyne DB	AIC	61D	36
Maskelyne F	IAU	LAC61	36
Maskelyne FD	AIC	61D	36
Maskelyne G	IAU	LAC60	35
Maskelyne GA	AIC	60C	35
Maskelyne HA	AIC	61D	36
Maskelyne J	IAU	LAC61	36
Maskelyne JA	AIC	61D	36
Maskelyne K	IAU	LAC60	36
Maskelyne KA	AIC	60C	36
Maskelyne KB	AIC	60C	36
Maskelyne KC	AIC	60C	36
Maskelyne M	IAU	LAC60	35, 36
Maskelyne MA	AIC	60C	35, 36
Maskelyne MB	AIC	60C	35, 36
Maskelyne N	IAU	LAC61	36
Maskelyne NA	AIC	60C	36
Maskelyne P	IAU	LAC61	36
Maskelyne PA	A&G	73-1	36
Maskelyne PB	AIC	61D	36
Maskelyne R	IAU	LAC61	36
Maskelyne T	IAU	LAC79	36
Maskelyne TA	AIC	61D	36
Maskelyne W	IAU	LAC60	36
Maskelyne X	IAU	LAC60	35, 36
Maskelyne Y	IAU	LAC60	36
Maskelyne α	AIC	61D	36
Maskelyne ε	AIC	61D	36
Maskelyne ζ	AIC	61D	36
Maskelyne η	AIC	61D	36

Feature	Source	Reference	Map(s)
Maskelyne θ	A&G	78-1	36
Maskelyne ι	AIC	61D	36
Maskelyne κ	AIC	61D	36
Maskelyne λ	AIC	61D	36
Maskelyne φ	A&G	78-1	36
Mason	IAU	LAC26	14
Mason A	IAU	LAC26	14
Mason B	IAU	LAC26	14
Mason C	IAU	LAC26	14
Matthew	CAF	B4	Detail 2
Maupertuis	IAU	LAC11	2
Maupertuis A	IAU	LAC11	2, 3
Maupertuis B	IAU	LAC11	2
Maupertuis C	IAU	LAC11	2, 3
Maupertuis K	IAU	LAC11	2
Maupertuis L	IAU	LAC11	2
Maury	IAU	LAC27	15
Maury A	IAU	LAC27	15
Maury AA	A&G	74-2	15
Maury B	IAU	LAC27	15
Maury C	IAU	LAC27	15
Maury D	IAU	LAC27	15
Maury J	IAU	LAC27	15
Maury K	IAU	LAC27	15
Maury L	IAU	LAC27	15
Maury M	IAU	LAC27	15
Maury N	IAU	LAC27	15
Maury P	IAU	LAC27	15
Maury T	IAU	LAC27	15
Maury U	IAU	LAC27	15
Mavis	IAU	LAC40	20
McDonald	IAU	LAC40	10
Menelaus	IAU	LAC42	23
Menelaus A	IAU	LAC42	23
Menelaus AB	A&G	90-2	23
Menelaus C	IAU	LAC60	23
Menelaus D	IAU	LAC60	34, 35
Menelaus E	IAU	LAC60	34
Menelaus R	A&G	90-2	23, 34
Menelaus α	A&G	90-2	23, 24, 34, 35
Menelaus ζ	NLC	LAC42	23
Menzel	IAU	LAC61	36
Mercurius	IAU	LAC28	15
Mercurius A	IAU	LAC28	7, 16
Mercurius B	IAU	LAC28	7, 15
Mercurius C	IAU	LAC28	7, 15
Mercurius CA	A&G	67-3	15
Mercurius D	IAU	LAC28	15, 16
Mercurius E	IAU	LAC15	7
Mercurius F	IAU	LAC28	15
Mercurius G	IAU	LAC28	15, 16
Mercurius H	IAU	LAC14	7
Mercurius J	IAU	LAC28	15
Mercurius K	IAU	LAC28	16
Mercurius L	IAU	LAC28	15

Feature	Source	Reference	Map(s)
Mercurius M	IAU	LAC15	7
Messala	[illegible]	[illegible]	[illegible]
Messala A	IAU	LAC27	16
Messala B	IAU	LAC28	16
Messala C	IAU	LAC28	16
Messala D	IAU	LAC28	16
Messala E	IAU	LAC28	16
Messala F	IAU	LAC28	16
Messala G	IAU	LAC28	16
Messala J	IAU	LAC28	16
Messala K	IAU	LAC28	16
Messier B	IAU	LAC79	37
Meton	IAU	LAC4	4, 5
Meton A	IAU	LAC4	5
Meton B	IAU	LAC4	4, 5
Meton C	IAU	LAC4	4, 5
Meton D	IAU	LAC4	5
Meton E	IAU	LAC4	4
Meton F	IAU	LAC4	4
Meton G	IAU	LAC4	5
Meton W	IAU	LAC4	5
Michael	IAU	LAC41	22
Milichius	IAU	LAC57	30
Milichius A	IAU	LAC57	30
Milichius B	A&G	133-2	30
Milichius BA	A&G	133-2	30
Milichius C	IAU	LAC58	30
Milichius D	IAU	LAC58	30
Milichius E	IAU	LAC58	30
Milichius K	IAU	LAC57	30
Milichius α	A&G	133-2	30
Milichius β	A&G	133-2	30, 31
Milichius κ	A&G	133-2	30
Milichius π	A&G	133-2	30
Milichius τ	A&G	138-2	30
Milichius ω	A&G	133-2	30
Misty Doublet	CAF	B4	Detail 2
Mitchell	IAU	LAC13	5, Detail 6
Mitchell A	A&G	103-3	5
Mitchell B	IAU	LAC13	5, 13
Mitchell E	IAU	LAC26	5, 13
MOCR	CAF	B6	Detail 6
Moigno	IAU	LAC4	5
Moigno A	IAU	LAC4	5
Moigno B	IAU	LAC4	5
Moigno C	IAU	LAC4	5
Moigno D	IAU	LAC4	5
Moltke	IAU	LAC78	35
Moltke A	IAU	LAC78	35
Moltke AC	A&G	85-1	35
Moltke AD	A&G	85-1	35
Moltke B	IAU	LAC78	35
Mons Agnes	IAU	LAC41	Detail 3
Mons Ampère	IAU	LAC41	22
Mons Argaeus	IAU	LAC42	24, 25
Mons Bradley	IAU	LAC41	22
Mons [illegible]	[illegible]	[illegible]	[illegible]
Mons Esam	IAU	LAC61	25, 36
Mons Gruithuisen Delta	IAU	LAC23	9
Mons Gruithuisen Gamma	IAU	LAC23	9
Mons Hadley	IAU	LAC41	22
Mons Hadley Delta	IAU	LAC41	22
Mons Heng	IAU	LAC23	9
Mons Herodotus	IAU	LAC38	18
Mons Hua	IAU	LAC23	9
Mons Huygens	IAU	LAC41	22
Mons La Hire	IAU	LAC40	20
Mons Latreille	IAU	LAC44	27
Mons Maraldi	IAU	LAC43	25
Mons Pico	IAU	LAC25	11
Mons Piton	IAU	LAC25	12
Mons Rümker	IAU	LAC23	8
Mons Usov	IAU	LAC62	38
Mons Vinogradov	IAU	LAC39	19
Mons Vitruvius	IAU	LAC43	25
Mons Wolff	IAU	LAC41	21
Mont Blanc	IAU	LAC25	12
Montes Agricola	IAU	LAC38	8, 18
Montes Alpes	IAU	LAC12	4, 12
Montes Apenninus	IAU	LAC58	21, 22, 23
Montes Archimedes	IAU	LAC41	21, 22
Montes Carpatus	IAU	LAC58	19, 20, 31
Montes Caucasus	IAU	LAC26	13
Montes Haemus	IAU	LAC41	23, 24
Montes Harbinger	IAU	LAC39	19
Montes Jura	IAU	LAC11	2, 10
Montes Recti	IAU	LAC12	3, 11
Montes Recti B	IAU	LAC12	3, 11
Montes Recti β	A&G	134-3	3
Montes Recti ϵ	A&G	134-3	3
Montes Secchi	IAU	LAC61	37
Montes Spitzbergen	IAU	LAC25	12
Montes Taurus	IAU	LAC43	25
Montes Teneriffe	IAU	LAC25	3, 11
Montes Teneriffe α	A&G	127-3	3
Montes Teneriffe δ	A&G	127-3	3
Montes Teneriffe ϵ	A&G	127-3	3
Montes Teneriffe ι	A&G	127-3	3
Montes Teneriffe ω	A&G	127-3	3
Mösting	IAU	LAC77	32
Mösting D	IAU	LAC76	32, 33
Mösting E	IAU	LAC59	33
Mösting K	IAU	LAC76	32
Mösting L	IAU	LAC76	33
Mösting δ	A&G	114-1	32
Mouchez	IAU	LAC3	3, 4
Mouchez A	IAU	LAC1	3, 4
Mouchez B	IAU	LAC3	4
Mouchez C	IAU	LAC3	3, 4

Feature	Source	Reference	Map(s)
Mouchez J	IAU	LAC2	3
Mouchez L	IAU	LAC2	3
Mouchez M	IAU	LAC1	3
Mount Marilyn	IAU	LAC61	36, 37
Murchison	IAU	LAC59	33
Murchison T	IAU	LAC59	33
N. Twin	CAF	B4	Detail 2
Nameless	CAF	B4	Detail 2
Nansen A	IAU	LAC1	5
Nansen C	IAU	LAC1	5
Nansen D	IAU	LAC1	5
Nansen-Apollo	IAU	LAC43	Detail 6
Natasha	IAU	LAC39	19
Naumann	IAU	LAC22	8
Naumann B	IAU	LAC23	8
Naumann G	IAU	LAC23	8
Neison	IAU	LAC4	5
Neison A	IAU	LAC4	5
Neison B	IAU	LAC4	5
Neison C	IAU	LAC4	5
Neison D	IAU	LAC4	5
Nemo	CAF	B6	Detail 6
Neper	IAU	LAC63	38
Neper D	IAU	LAC63	38
Neper H	IAU	LAC63	38
Neper Q	IAU	LAC63	38
Newcomb	IAU	LAC43	15, 25
Newcomb A	IAU	LAC43	25
Newcomb B	IAU	LAC43	25
Newcomb C	IAU	LAC43	15, 25
Newcomb F	IAU	LAC43	15
Newcomb G	IAU	LAC43	25
Newcomb H	IAU	LAC43	25
Newcomb J	IAU	LAC43	25
Newcomb P	A&G	66-3	25, 26
Newcomb PA	A&G	66-3	26
Newcomb Q	IAU	LAC43	15
Nielsen	IAU	LAC38	8
Nikolya	IAU	LAC24	Detail 1
Nobili	IAU	LAC63	38
North Complex	IAU	LAC41	Detail 2
November	CAF	B4	Detail 2
Oceanus Procellarum	IAU	LAC10	1, 8, 17, 18, 28, 29
Oenopides	IAU	LAC10	1
Oenopides B	IAU	LAC10	1
Oenopides K	IAU	LAC10	1
Oenopides L	IAU	LAC10	1
Oenopides M	IAU	LAC10	1
Oenopides R	IAU	LAC10	1
Oenopides S	IAU	LAC10	1
Oenopides T	IAU	LAC10	1
Oenopides X	IAU	LAC10	1
Oenopides Y	IAU	LAC10	1
Oenopides Z	IAU	LAC10	1
Oersted	IAU	LAC27	15
Oersted A	IAU	LAC27	15
Oersted P	IAU	LAC27	15
Oersted U	IAU	LAC27	15
Offset Crater	CAF	B4	Detail 2
Olbers	IAU	LAC55	28
Olbers B	IAU	LAC55	28
Olbers D	IAU	LAC55	28
Olbers G	IAU	LAC55	28
Olbers H	IAU	LAC55	28
Olbers K	IAU	LAC55	28
Olbers M	IAU	LAC55	28
Olbers N	IAU	LAC55	28
Olbers S	IAU	LAC55	28
Olbers V	IAU	LAC55	28
Olbers W	IAU	LAC55	28
Olbers Y	IAU	LAC55	28
Oppolzer	IAU	LAC77	33
Oppolzer A	IAU	LAC77	33
Orville	CAF	B4	Detail 2
Os	CAF	B4	Detail 2
Osama	IAU	LAC41	Detail 3
Osiris	IAU	LAC42	Detail 5
Pallas	IAU	LAC59	33
Pallas A	IAU	LAC59	33
Pallas B	IAU	LAC59	33
Pallas C	IAU	LAC59	33
Pallas D	IAU	LAC59	33
Pallas E	IAU	LAC59	33
Pallas F	IAU	LAC59	33
Pallas FA	AIC	59D	33
Pallas H	IAU	LAC59	33
Pallas N	IAU	LAC59	33
Pallas V	IAU	LAC59	33
Pallas W	IAU	LAC59	33
Pallas X	IAU	LAC59	33
Pallas η	A&G	109-1	33
Pallas ϕ	A&G	109-1	33
Pallas ω	A&G	109-1	33
Palus Nebularum	LC	12-21	12
Palus Putredinis	IAU	LAC41	22
Palus Somni	IAU	LAC61	37
Pascal	IAU	LAC2	2, 3
Pascal A	IAU	LAC2	2
Pascal F	IAU	LAC2	3
Pascal G	IAU	LAC2	2, 3
Pascal J	IAU	LAC2	2
Pascal L	IAU	LAC2	3
Patricia	IAU	LAC41	22
Peek	IAU	LAC63	38
Pei Xiu	IAU	LAC23	9
Peirce	IAU	LAC44	26
Peirce C	IAU	LAC43	26
Petermann	IAU	LAC5	5
Petermann B	IAU	LAC5	5

Feature	Source	Reference	Map(s)
Petermann C	IAU	LAC5	5
Petermann D	IAU	[illegible]	[illegible]
Petermann E	IAU	LAC4	5
Petermann R	IAU	LAC4	5
Petermann S	IAU	LAC5	5
Peters	IAU	LAC4	5
Petit	IAU	LAC62	38
Philolaus	IAU	LAC3	3
Philolaus B	IAU	LAC3	3
Philolaus C	IAU	LAC3	3
Philolaus D	IAU	LAC3	3
Philolaus E	IAU	LAC3	3
Philolaus F	IAU	LAC3	3
Philolaus G	IAU	LAC3	3
Philolaus U	IAU	LAC3	3
Philolaus W	IAU	LAC2	3
Piazzi Smyth	IAU	LAC25	12
Piazzi Smyth B	IAU	LAC25	12
Piazzi Smyth M	IAU	LAC25	12
Piazzi Smyth U	IAU	LAC25	12
Piazzi Smyth V	IAU	LAC25	12
Piazzi Smyth W	IAU	LAC25	12
Piazzi Smyth Y	IAU	LAC25	12
Piazzi Smyth Z	IAU	LAC25	12
Piazzi Smyth α	A&G	115-2	12
Piazzi Smyth β	A&G	115-2	12
Piazzi Smyth π	A&G	115-2	12
Picard	IAU	LAC62	26, 37
Picard K	IAU	LAC62	37
Picard L	IAU	LAC62	37
Picard M	IAU	LAC62	37
Picard N	IAU	LAC62	37
Picard P	IAU	LAC62	37
Picard Y	IAU	LAC62	38
Pico B	IAU	LAC24	11
Pico BA	A&G	127-2	11
Pico C	IAU	LAC25	12
Pico D	IAU	LAC25	11
Pico E	IAU	LAC25	11
Pico EA	A&G	127-2	11
Pico F	IAU	LAC25	11
Pico G	IAU	LAC25	11
Pico K	IAU	LAC25	11, 12
Pico β	A&G	122-2	11
Pitane	CAF	B4	Detail 2
Piton A	IAU	LAC25	12
Piton B	IAU	LAC25	12
Piton γ	A&G	110-1	12
Plain	IAU	LAC41	Detail 2
Plana	IAU	LAC26	14
Plana C	IAU	LAC26	14
Plana D	IAU	LAC26	14
Plana E	IAU	LAC26	14
Plana F	IAU	LAC26	14
Plana G	IAU	LAC26	14
Plana β	A&G	91-2	14
Plana γ	[illegible]	[illegible]	[illegible]
Plana δ	A&G	91-2	14
Plana ϵ	A&G	91-2	14
Plana ζ	A&G	91-2	14
Plana ξ	A&G	91-2	14
Planitia Descensus	IAU	LAC56	28
Plato	IAU	LAC12	3, 4
Plato B	IAU	LAC12	3
Plato BA	A&G	128-1	3
Plato BB	A&G	140-1	3
Plato C	IAU	LAC12	3
Plato D	IAU	LAC12	3
Plato E	IAU	LAC12	3
Plato F	IAU	LAC12	3
Plato G	IAU	LAC12	4
Plato H	IAU	LAC12	4
Plato HA	A&G	116-1	4
Plato J	IAU	LAC12	4
Plato K	IAU	LAC25	12
Plato KA	IAU	LAC25	12
Plato KB	A&G	115-3	12
Plato L	IAU	LAC12	4
Plato M	IAU	LAC12	3
Plato O	IAU	LAC12	3
Plato P	IAU	LAC12	3
Plato Q	IAU	LAC12	4
Plato R	IAU	LAC12	3
Plato S	IAU	LAC12	3
Plato T	IAU	LAC12	3
Plato U	IAU	LAC12	4
Plato V	IAU	LAC12	4
Plato VA	A&G	128-1	3, 4
Plato W	IAU	LAC12	3
Plato X	IAU	LAC12	3
Plato Y	IAU	LAC12	3
Plato Z	A&G	128-1	3
Plato ζ	A&G	128-1	3
Plato η	A&G	140-1	3
Plato κ	A&G	116-1	4
Plato ν	A&G	128-1	3, 4
Plato π	A&G	116-1	4
Plato ρ	A&G	140-1	3
Plato σ	A&G	128-1	3, 4
Plato υ	A&G	116-1	4
Plato ϕ	A&G	128-1	3
Plato χ	A&G	116-1	4
Plato ω	A&G	116-1	4
Plinius	IAU	LAC60	24
Plinius A	IAU	LAC60	35
Plinius B	IAU	LAC60	24, 35
Plutarch	IAU	LAC45	27
Plutarch C	IAU	LAC45	27
Plutarch CA	NLC	LAC44	27
Plutarch D	IAU	LAC45	27

Feature	Source	Reference	Map(s)
Plutarch F	IAU	LAC45	27
Plutarch G	IAU	LAC45	27
Plutarch H	IAU	LAC45	27
Plutarch K	IAU	LAC45	27
Plutarch L	IAU	LAC45	27
Plutarch M	IAU	LAC45	27
Plutarch N	IAU	LAC45	27
Pluton	CAF	B4	Detail 2
Pomortsev	IAU	LAC62	38
Poncelet	IAU	LAC2	3
Poncelet B	IAU	LAC2	3
Poncelet C	IAU	LAC2	3
Poncelet D	IAU	LAC2	3
Poncelet H	IAU	LAC2	3
Poncelet P	IAU	LAC1	3
Poncelet Q	IAU	LAC2	3
Poncelet R	IAU	LAC2	3
Poncelet S	IAU	LAC2	3
Pooh	CAF	B4	Detail 2
Poppie	CAF	B6	Detail 6
Posidonius	IAU	LAC42	14
Posidonius A	IAU	LAC42	14
Posidonius B	IAU	LAC26	14
Posidonius C	IAU	LAC42	14
Posidonius D	A&G	86-1	14
Posidonius E	IAU	LAC42	14
Posidonius F	IAU	LAC26	14
Posidonius FA	A&G	86-1	14
Posidonius G	IAU	LAC26	14
Posidonius J	IAU	LAC26	14
Posidonius M	IAU	LAC26	14
Posidonius N	IAU	LAC42	14, 24
Posidonius NA	A&G	86-1	24
Posidonius NB	A&G	91-1	24
Posidonius NC	A&G	86-1	24
Posidonius ND	A&G	86-1	24
Posidonius NE	A&G	86-1	24
Posidonius O	A&G	86-1	14
Posidonius P	IAU	LAC26	14
Posidonius V	NLC	LAC42	14
Posidonius W	IAU	LAC42	14
Posidonius Y	IAU	LAC42	14
Posidonius Z	IAU	LAC42	14
Powel	CAF	B6	Detail 6
Prinz	IAU	LAC39	18, 19
Proclus	IAU	LAC43	26
Proclus A	IAU	LAC61	37
Proclus C	IAU	LAC61	37
Proclus D	IAU	LAC43	25
Proclus E	IAU	LAC43	25
Proclus G	IAU	LAC61	37
Proclus GA	A&G	66-2	37
Proclus J	IAU	LAC43	26
Proclus K	IAU	LAC43	26
Proclus L	IAU	LAC43	26

Feature	Source	Reference	Map(s)
Proclus M	IAU	LAC43	26
Proclus P	IAU	LAC61	26
Proclus PA	A&G	61-2	26
Proclus R	IAU	LAC43	26
Proclus S	IAU	LAC61	26
Proclus T	IAU	LAC61	26
Proclus U	IAU	LAC61	26
Proclus V	IAU	LAC61	26, 37
Proclus W	IAU	LAC43	26
Proclus X	IAU	LAC43	26
Proclus Y	IAU	LAC43	26
Proclus Z	IAU	LAC43	26
Proclus ξ	A&G	61-2	26
Prom La Hire C	A&G	133-3	20
Prom La Hire α	A&G	133-3	20
Promontorium Agarum	IAU	LAC62	27, 38
Promontorium Agassiz	IAU	LAC25	12
Promontorium Archerusia	IAU	LAC42	24
Promontorium Deville	IAU	LAC25	12
Promontorium Fresnel	IAU	LAC41	22
Promontorium Heraclides	IAU	LAC24	10
Promontorium Laplace	IAU	LAC24	10
Promontorium Lavinium	A&G	61-2	26
Promontorium Olivium	A&G	61-2	26
Protagoras	IAU	LAC12	4
Protagoras B	IAU	LAC12	4
Protagoras E	IAU	LAC12	4
Protagoras ϵ	A&G	103-3	5
Punk	CAF	B6	Detail 6
Pupin	IAU	LAC40	21
Pythagoras	IAU	LAC10	2
Pythagoras B	IAU	LAC2	2
Pythagoras D	IAU	LAC2	2
Pythagoras E	A&G	164-1	2
Pythagoras G	IAU	LAC2	2
Pythagoras H	IAU	LAC2	2
Pythagoras K	IAU	LAC2	2
Pythagoras L	IAU	LAC2	2
Pythagoras M	IAU	LAC9	2
Pythagoras N	IAU	LAC2	2
Pythagoras P	IAU	LAC2	2
Pythagoras S	IAU	LAC2	2
Pythagoras T	IAU	LAC10	2
Pythagoras W	IAU	LAC11	2
Pythagoras α	A&G	190-2	2
Pytheas	IAU	LAC40	20
Pytheas A	IAU	LAC40	20
Pytheas B	IAU	LAC40	20
Pytheas C	IAU	LAC40	20
Pytheas D	IAU	LAC40	20
Pytheas E	IAU	LAC40	20
Pytheas F	IAU	LAC40	20
Pytheas G	IAU	LAC40	20
Pytheas H	IAU	LAC40	20, 21

Feature	Source	Reference	Map(s)
Pytheas J	IAU	LAC40	20
Pytheas K	IAU	[illegible]	[illegible]
Pytheas L	IAU	LAC40	20, 21
Pytheas M	IAU	LAC40	20
Pytheas N	IAU	LAC40	20
Pytheas U	IAU	LAC40	20
Pytheas W	IAU	LAC40	20
Pytheas β	A&G	126-3	20
Quadrant	CAF	B4	Detail 2
Quark	CAF	B4	Detail 2
Raman	IAU	LAC38	18
Rayleigh C	IAU	LAC45	16
Réaumur D	IAU	LAC77	33
Recti β	A&G	134-3	11
Recti ϵ	A&G	134-3	11
Reiner	IAU	LAC56	28, 29
Reiner A	IAU	LAC56	29
Reiner C	IAU	LAC56	29
Reiner E	IAU	LAC57	29
Reiner G	IAU	LAC56	28, 29
Reiner Gamma	IAU	LAC56	28
Reiner H	IAU	LAC56	28, 29
Reiner K	IAU	LAC56	29
Reiner L	IAU	LAC56	28, 29
Reiner M	IAU	LAC56	28
Reiner N	IAU	LAC56	28
Reiner P	A&G	157-1	29
Reiner Q	IAU	LAC56	29
Reiner R	IAU	LAC56	28
Reiner S	IAU	LAC56	29
Reiner T	IAU	LAC56	29
Reiner U	IAU	LAC56	29
Reinhold	IAU	LAC58	31
Reinhold A	IAU	LAC58	31
Reinhold B	IAU	LAC58	31
Reinhold C	IAU	LAC58	31
Reinhold D	IAU	LAC58	31
Reinhold F	IAU	LAC58	31
Reinhold G	IAU	LAC58	31
Reinhold H	IAU	LAC58	31
Reinhold N	IAU	LAC58	31
Reinhold NA	A&G	126-1	31
Reinhold β	A&G	126-1	31
Reinhold γ	A&G	126-1	31
Reinhold δ	NLC	LAC58	31
Reinhold θ	A&G	126-1	31
Reinhold ι	A&G	126-1	31
Reinhold χ	A&G	126-1	31
Repsold	IAU	LAC10	1
Repsold A	IAU	LAC10	1
Repsold B	IAU	LAC10	1
Repsold C	IAU	LAC10	1, 8
Repsold G	IAU	LAC21	1
Repsold H	IAU	LAC21	1
Repsold R	IAU	LAC10	1
Repsold S	IAU	LAC22	1, 8
[illegible]	[illegible]	[illegible]	[illegible]
Repsold V	IAU	LAC10	1
Repsold W	IAU	LAC21	1
Respighi	IAU	LAC63	38
Rhaeticus	IAU	LAC59	33, 34
Rhaeticus A	IAU	LAC59	33, 34
Rhaeticus B	IAU	LAC59	34
Rhaeticus D	IAU	LAC59	34
Rhaeticus DA	A&G	97-1	34
Rhaeticus E	IAU	LAC77	34
Rhaeticus F	IAU	LAC59	34
Rhaeticus G	IAU	LAC59	34
Rhaeticus H	IAU	LAC77	33, 34
Rhaeticus J	IAU	LAC77	33
Rhaeticus L	IAU	LAC59	33
Rhaeticus M	IAU	LAC59	33
Rhaeticus N	IAU	LAC59	33
Rhysling	IAU	LAC41	Detail 2
Riccioli C	IAU	LAC55	28
Riccioli CA	IAU	LAC55	28
Riccioli G	IAU	LAC73	28
Riccioli H	IAU	LAC55	28
Ridge	CAF	B4	Detail 2
Riemann B	IAU	LAC29	16
Rim	CAF	B4	Detail 2
Rima Agricola	IAU	LAC38	8, 18
Rima Archimedes I	A&G	115-1	21
Rima Archimedes II	A&G	115-1	21
Rima Archimedes III	A&G	109-3	22
Rima Archimedes IV	A&G	109-3	22
Rima Archimedes V	A&G	109-3	22
Rima Archimedes VI	A&G	109-3	22
Rima Archytas	IAU	LAC12	4
Rima Archytas I	A&G	116-1	4
Rima Ariadaeus	IAU	LAC60	34, 35
Rima Ariadaeus I	A&G	90-1	35
Rima Aristarchus I	A&G	151-1	18
Rima Aristarchus II	A&G	151-1	18
Rima Aristarchus III	A&G	151-1	18, 19
Rima Aristarchus IV	A&G	151-1	18, 19
Rima Aristarchus V	A&G	151-1	18, 19
Rima Aristarchus VI	A&G	151-1	18, 19
Rima Aristarchus VII	A&G	151-1	18
Rima Aristarchus VIII	A&G	151-1	19
Rima Aristoteles I	A&G	103-3	5
Rima Aristoteles II	A&G	103-3	5
Rima Artsimovich	IAU	LAC39	19
Rima Atlas I	A&G	74-2	15
Rima Atlas II	A&G	74-2	15
Rima Atlas III	A&G	74-2	15
Rima Atlas IV	A&G	74-2	15
Rima Atlas V	A&G	74-2	15
Rima Bode I	A&G	109-2	33
Rima Bode II	A&G	109-2	33

Feature	Source	Reference	Map(s)
Rima Bode III	A&G	109-1	33
Rima Bode IV	A&G	109-1	32, 33
Rima Boscovich I	A&G	97-2	34
Rima Boscovich II	A&G	97-2	34
Rima Bradley	IAU	LAC41	22
Rima Brayley	IAU	LAC39	19
Rima Burckhardt I	A&G	62-1	16
Rima Calippus	IAU	LAC26	13
Rima Calippus I	A&G	98-2	13
Rima Cardanus	IAU	LAC55	28
Rima Carmen	IAU	LAC42	24, 25
Rima Cauchy	IAU	LAC61	36
Rima Cauchy I	A&G	73-2	36
Rima Chacornac I	A&G	86-1	14, 15, 24
Rima Chacornac II	A&G	79-1	15
Rima Chacornac IV	A&G	86-2	14, 24
Rima Cleomedes	IAU	LAC44	26
Rima Cleomedes I	A&G	54-3	26
Rima Cleopatra	IAU	LAC38	8
Rima Conon	IAU	LAC41	22
Rima Daniell I	A&G	86-2	14
Rima Daniell III	A&G	91-2	14
Rima Daniell IV	A&G	91-2	14
Rima Dawes	IAU	LAC42	24
Rima Delisle	IAU	LAC39	9
Rima Diophantus	IAU	LAC39	19
Rima Draper	IAU	LAC40	20
Rima Eudoxus I	A&G	103-2	13
Rima Eudoxus II	A&G	103-2	13
Rima Euler	IAU	LAC39	19
Rima Fresnel I	A&G	102-3	22
Rima Fresnel II	A&G	102-3	22
Rima Fresnel III	A&G	102-3	22
Rima G. Bond	IAU	LAC27	15
Rima G. Bond I	A&G	79-1	15
Rima Galilaei	IAU	LAC56	28
Rima Gärtner	IAU	LAC13	6
Rima Gärtner I	A&G	86-3	6
Rima Gay-Lussac	IAU	LAC58	31
Rima Gay-Lussac I	A&G	126-2	31
Rima Gay-Lussac II	A&G	126-2	31
Rima Hadley	IAU	LAC41	22
Rima Hevelius I	A&G	169-1	28
Rima Hevelius II	A&G	169-1	28
Rima Hevelius III	A&G	169-1	28
Rima Hyginus	IAU	LAC59	33, 34
Rima Hyginus I	A&G	97-1	34
Rima Hypatia I	A&G	85-1	35
Rima Hypatia II	A&G	85-1	35
Rima Jansen	IAU	LAC60	25, 36
Rima Jansen I	A&G	78-2	36
Rima Krieger	IAU	LAC39	18
Rima Littrow I	A&G	78-3	24, 25
Rima Littrow II	A&G	78-3	25
Rima Littrow III	A&G	78-3	24, 25

Feature	Source	Reference	Map(s)
Rima Littrow IV	A&G	78-3	24, 25
Rima Littrow V	A&G	78-3	24, 25
Rima Littrow VI	A&G	78-3	24, 25
Rima Littrow VII	A&G	78-3	24, 25
Rima Louville	IAU	LAC23	9
Rima Mairan	IAU	LAC23	9
Rima Marcello	IAU	LAC42	Detail 5
Rima Marius	IAU	LAC38	18
Rima Maskelyne I	AIC	60C	35, 36
Rima Menelaus I	A&G	90-3	24
Rima Menelaus II	A&G	90-3	24
Rima Menelaus III	A&G	90-3	24
Rima Messier	IAU	LAC79	37
Rima Milichius	IAU	LAC57	30
Rima Newcomb I	A&G	74-1	15
Rima Plato I	A&G	116-1	4
Rima Plato II	A&G	116-1	4
Rima Plato III	A&G	140-1	3
Rima Plinius I	A&G	85-2	24
Rima Plinius II	A&G	85-2	24
Rima Plinius III	A&G	85-2	24
Rima Posidonius I	A&G	86-1	14
Rima Posidonius II	A&G	86-1	14
Rima Posidonius IV	A&G	86-1	14
Rima Posidonius V	A&G	86-1	14
Rima Posidonius VI	A&G	86-1	14
Rima Prinz I	A&G	151-1	19
Rima Prinz II	A&G	151-1	19
Rima Reiko	IAU	LAC42	Detail 5
Rima Rhaeticus I	A&G	102-1	33
Rima Riccioli I	A&G	169-1	28
Rima Ritter I	A&G	90-1	35
Rima Ritter II	A&G	90-1	35
Rima Ritter III	A&G	90-1	35
Rima Ritter V	A&G	90-1	35
Rima Römer I	A&G	78-3	25
Rima Römer II	A&G	79-1	15, 25
Rima Rudolf	IAU	LAC42	24, 25
Rima Schröter	IAU	LAC59	32
Rima Schröter I	A&G	109-1	32
Rima Schröter V	A&G	109-1	32
Rima Sharp	IAU	LAC11	1, 9
Rima Sharp I	A&G	158-2	9
Rima Sheepshanks	IAU	LAC13	5
Rima Sosigenes I	A&G	90-2	35
Rima Sosigenes II	A&G	90-2	35
Rima Sosigenes III	A&G	90-1	35
Rima Sosigenes IV	A&G	90-1	36
Rima Stadius I	A&G	121-2	20
Rima Stadius II	A&G	121-2	20, 31
Rima Suess	IAU	LAC57	29
Rima Sulpicius Gallus I	A&G	97-3	23
Rima Sulpicius Gallus II	A&G	97-3	23
Rima Sulpicius Gallus III	A&G	97-3	23
Rima Sung-Mei	IAU	LAC42	Detail 4

Feature	Source	Reference	Map(s)
Rima T. Mayer	IAU	LAC57	19, 30
Rima Theaetetus I	A&G	103-1	12
Rima Theaetetus II	A&G	103-1	12
Rima Theaetetus III	A&G	103-1	12
Rima Triesnecker I	A&G	102-1	33
Rima Triesnecker II	A&G	102-1	33, 34
Rima Triesnecker III	A&G	102-1	33, 34
Rima Triesnecker V	A&G	102-1	33
Rima Triesnecker VI	A&G	102-1	33, 34
Rima Triesnecker VII	A&G	102-1	33
Rima Vladimir	IAU	LAC42	22
Rima Wan-Yu	IAU	LAC39	19
Rima Yangel'	IAU	LAC41	22
Rima Zahia	IAU	LAC39	19
Rimae Apollonius	IAU	LAC62	37
Rimae Archimedes	IAU	LAC41	21, 22
Rimae Aristarchus	IAU	LAC39	8, 18, 19
Rimae Atlas	IAU	LAC27	15
Rimae Bode	IAU	LAC59	33
Rimae Boscovich	IAU	LAC60	34
Rimae Bürg	IAU	LAC26	14
Rimae Chacornac	IAU	LAC43	14, 15, 25
Rimae Daniell	IAU	LAC26	14
Rimae Fresnel	IAU	LAC41	22
Rimae Hevelius	IAU	LAC56	28
Rimae Hypatia	IAU	LAC78	35
Rimae Littrow	IAU	LAC42	24, 25
Rimae Maclear	IAU	LAC60	35
Rimae Maestlin	IAU	LAC57	29
Rimae Maupertuis	IAU	LAC11	3
Rimae Menelaus	IAU	LAC42	23, 24
Rimae Plato	IAU	LAC12	4
Rimae Plinius	IAU	LAC42	24
Rimae Posidonius	IAU	LAC26	14
Rimae Prinz	IAU	LAC39	19
Rimae Ritter	IAU	LAC60	35
Rimae Römer	IAU	LAC43	25
Rimae Secchi	IAU	LAC61	37
Rimae Sosigenes	IAU	LAC60	35
Rimae Sulpicius Gallus	IAU	LAC42	23
Rimae Taruntius	IAU	LAC61	37
Rimae Theaetetus	IAU	LAC25	12
Rimae Triesnecker	IAU	LAC59	33, 34
Ring	CAF	B4	Detail 2
Rio	CAF	B4	Detail 2
Ritter	IAU	LAC60	35
Ritter B	IAU	LAC60	35
Ritter C	IAU	LAC60	35
Ritter D	IAU	LAC60	35
Robert	IAU	LAC42	Detail 5
Robinson	IAU	LAC11	2
Rocco	IAU	LAC39	19
Rogers	CAF	B6	Detail 6
Roman Steppe	CAF	B6	Detail 6
Römer	IAU	LAC43	25

Feature	Source	Reference	Map(s)
Römer A	IAU	LAC43	25
Römer B	IAU	LAC43	25
Römer BA	A&G	73-3	25
Römer C	IAU	LAC43	25
Römer D	IAU	LAC43	25
Römer E	IAU	LAC43	25
Römer F	IAU	LAC43	25
Römer G	IAU	LAC43	25
Römer H	IAU	LAC43	25
Römer J	IAU	LAC43	25
Römer KA	A&G	73-3	25
Römer M	IAU	LAC43	25
Römer N	IAU	LAC43	25
Römer P	IAU	LAC43	25
Römer PA	A&G	73-3	25
Römer PB	A&G	73-3	25
Römer PC	A&G	73-3	25
Römer R	IAU	LAC43	25
Römer S	IAU	LAC43	25
Römer T	IAU	LAC43	25
Römer TA	A&G	73-3	25
Römer U	IAU	LAC43	25
Römer V	IAU	LAC43	25
Römer W	IAU	LAC43	25
Römer X	IAU	LAC43	25
Römer Y	IAU	LAC43	25
Römer Z	IAU	LAC43	25
Rondone Massif	CAF	B6	Detail 6
Rosa	IAU	LAC39	19
Ross	IAU	LAC60	35
Ross B	IAU	LAC60	35
Ross C	IAU	LAC60	35
Ross D	IAU	LAC60	35
Ross E	IAU	LAC60	35
Ross F	IAU	LAC60	35
Ross G	IAU	LAC60	35
Ross H	IAU	LAC60	35
Ross μ	A&G	85-2	35
Rudolph	CAF	B6	Detail 6
Ruin Basin	CAF	B1	Detail 8
Rümker C	IAU	LAC23	8, 9
Rümker E	IAU	LAC23	8
Rümker F	IAU	LAC23	8
Rümker H	IAU	LAC23	9
Rümker K	IAU	LAC23	9
Rümker L	IAU	LAC23	9
Rümker S	IAU	LAC22	8
Rümker T	IAU	LAC22	8
Rümker α	A&G	170-2	8, 9
Rümker β	A&G	170-5	8
Rümker ζ	A&G	170-3	8, 9
Rümker η	A&G	158-2	9
Rümker θ	A&G	170-4	8, 9
Rümker ξ	A&G	158-2	9
Rupes Boris	IAU	LAC39	9

Feature	Source	Reference	Map(s)
Rupes Cauchy	IAU	LAC61	36
Rupes Toscanelli	IAU	LAC39	18
Russell	IAU	LAC37	17
Russell B	IAU	LAC37	17
Russell E	IAU	LAC37	17
Russell F	IAU	LAC37	17
Russell R	IAU	LAC37	17
Russell S	IAU	LAC37	17
Ruth	IAU	LAC39	19
S P Crater	CAF	B1	Detail 8
S. Twin	CAF	B4	Detail 2
Sabatier	IAU	LAC63	38
Sabine	IAU	LAC60	35
Sabine A	IAU	LAC60	35
Sabine AB	AIC	60D	35
Sabine AC	AIC	60D	35
Sabine AD	A&G	85-1	35
Sabine C	IAU	LAC60	35
Sabine CA	A&G	85-1	35
Sabine DM	AIC	60C	35
Sabine EA	AIC	60C	35
Sabine EB	AIC	60C	35
Sabine EF	AIC	60C	35
Sabine α	AIC	60D	35
Sabine β	AIC	60D	35
Salyut	CAF	B4	Detail 2
Samir	IAU	LAC39	19
Sampson	IAU	LAC40	11, 21
Samstag	CAF	B4	Detail 2
San Luis Ray	CAF	B6	Detail 6
Santos-Dumont	IAU	LAC41	22
Sarabhai	IAU	LAC42	24
Scarp	CAF	B4	Detail 2, Detail 6
Schiaparelli	IAU	LAC38	18
Schiaparelli A	IAU	LAC38	17, 18
Schiaparelli C	IAU	LAC38	18
Schiaparelli E	IAU	LAC38	18
Schmidt	IAU	LAC60	35
Schmidt A	A&G	85-1	35
Schröter	IAU	LAC59	32
Schröter A	IAU	LAC59	32
Schröter C	IAU	LAC59	32
Schröter D	IAU	LAC59	32
Schröter E	IAU	LAC59	32
Schröter F	IAU	LAC59	32
Schröter FA	A&G	109-1	32
Schröter G	IAU	LAC59	32
Schröter H	IAU	LAC59	32
Schröter J	IAU	LAC59	32
Schröter K	IAU	LAC59	32
Schröter KA	A&G	109-1	32
Schröter KB	A&G	109-1	32
Schröter L	IAU	LAC59	32
Schröter M	IAU	LAC58	32
Schröter S	IAU	LAC59	32

Feature	Source	Reference	Map(s)
Schröter T	IAU	LAC59	32
Schröter U	IAU	LAC59	32
Schröter W	IAU	LAC59	32
Schubert	IAU	LAC63	38
Schubert A	IAU	LAC63	38
Schubert C	IAU	LAC63	38
Schubert E	IAU	LAC63	38
Schubert F	IAU	LAC63	38
Schubert G	IAU	LAC63	38
Schubert H	IAU	LAC63	38
Schubert J	IAU	LAC63	38
Schubert K	IAU	LAC63	38
Schubert N	IAU	LAC63	38
Schubert X	IAU	LAC63	38
Schumacher	IAU	LAC28	15, 16
Schumacher B	IAU	LAC28	15, 16
Schwabe	IAU	LAC4	6
Schwabe C	IAU	LAC4	5, 6
Schwabe D	IAU	LAC4	6
Schwabe E	IAU	LAC4	6
Schwabe F	IAU	LAC4	6
Schwabe G	IAU	LAC4	5, 6
Schwabe K	IAU	LAC4	6
Schwabe N	A&G	80-2	5
Schwabe U	IAU	LAC5	6
Schwabe W	IAU	LAC4	5, 6
Schwabe X	IAU	LAC5	6
Scoresby	IAU	LAC4	4
Scoresby AA	A&G	116-3	4
Scoresby K	IAU	LAC3	4
Scoresby M	IAU	LAC3	4
Scoresby P	IAU	LAC4	4
Scoresby Q	IAU	LAC3	4
Scoresby W	IAU	LAC4	4
Sculptured Hills	IAU	LAC43	Detail 6
Secchi	IAU	LAC61	37
Secchi A	IAU	LAC61	37
Secchi B	IAU	LAC61	37
Secchi G	IAU	LAC61	37
Secchi K	IAU	LAC79	37
Secchi O	CAF	B1	Detail 8
Secchi U	IAU	LAC61	37
Secchi UA	A&G	66-1	37, Detail 8
Secchi UB	A&G	66-1	37
Secchi X	IAU	LAC79	37
Secchi α	AIC	61D	36
Secchi β	AIC	61D	36
Seleucus	IAU	LAC38	17
Seleucus A	IAU	LAC38	17, 18
Seleucus E	IAU	LAC38	17
Seneca	IAU	LAC45	27
Seneca A	IAU	LAC45	27
Seneca B	IAU	LAC45	27
Seneca C	IAU	LAC45	27
Seneca D	IAU	LAC45	27

Feature	Source	Reference	Map(s)
Seneca F	IAU	LAC45	16
[illegible]	[illegible]	[illegible]	10
Shakespeare	IAU	LAC43	Detail 6
Shapley	IAU	LAC62	37, 38
Sharp	IAU	LAC23	9, 10
Sharp A	IAU	LAC23	1, 2, 9, 10
Sharp B	IAU	LAC23	9
Sharp C	A&G	158-2	9, 10
Sharp D	IAU	LAC23	9
Sharp J	IAU	LAC23	10
Sharp K	IAU	LAC23	2, 10
Sharp L	IAU	LAC23	10
Sharp M	IAU	LAC23	2, 9, 10
Sharp U	IAU	LAC23	1, 9
Sharp V	IAU	LAC23	9
Sharp W	IAU	LAC11	1, 2
Sharp η	A&G	163-3	1
Sharp ξ	A&G	163-3	1, 2
Sheepshanks	IAU	LAC13	5
Sheepshanks A	IAU	LAC13	5
Sheepshanks B	IAU	LAC13	5
Sheepshanks C	IAU	LAC13	5
Shen Kuo	IAU	LAC23	9
Sherlock	IAU	LAC43	Detail 6
Shorty	IAU	LAC43	Detail 6
Shuckburgh	IAU	LAC27	15
Shuckburgh A	IAU	LAC27	15
Shuckburgh C	IAU	LAC27	15
Shuckburgh E	IAU	LAC27	15
Side	CAF	B4	Detail 2
Sidewinder Rille	CAF	B1	Detail 8
Silberschlag	IAU	LAC60	34
Silberschlag A	IAU	LAC60	34
Silberschlag AB	AIC	60D	34
Silberschlag D	IAU	LAC60	34
Silberschlag E	IAU	LAC60	34
Silberschlag G	IAU	LAC60	34
Silberschlag P	IAU	LAC60	34
Silberschlag S	IAU	LAC60	34
Silberschlag β	AIC	60D	34
Sinas	IAU	LAC61	36
Sinas A	IAU	LAC61	36
Sinas E	IAU	LAC61	36
Sinas G	IAU	LAC61	36
Sinas H	IAU	LAC61	36
Sinas J	IAU	LAC61	36
Sinas K	IAU	LAC61	36
Sinas α	A&G	78-1	36
Sinas β	A&G	78-2	36
Sinus Aestuum	IAU	LAC59	21, 32
Sinus Amoris	IAU	LAC43	25
Sinus Concordiae	IAU	LAC61	37
Sinus Fidei	IAU	LAC41	22
Sinus Honoris	IAU	LAC60	35
Sinus Iridum	IAU	LAC24	10
Sinus Lunicus	IAU	LAC25	12
[illegible]	[illegible]	[illegible]	[illegible]
Sinus Roris	IAU	LAC10	1, 2
Sinus Successus	IAU	LAC62	38
Sinus Viscositatis	IAU	LAC23	9
Slava	IAU	LAC24	Detail 1
Slide Crater	CAF	B4	Detail 2
Smith	IAU	B6	Detail 6
Smithson	IAU	LAC62	37
Smokey Basin	CAF	B1	Detail 8
Smokey Valley	CAF	B1	Detail 8
Snake Ridge	CAF	B1	Detail 8
Snoopy	CAF	B6	Detail 6
Sömmering	IAU	LAC59	32
Sömmering A	IAU	LAC58	32
Sömmering P	IAU	LAC58	32
Sömmering R	IAU	LAC59	32
Sömmering β	A&G	109-1	32
Sömmering γ	A&G	109-1	32
Sömmering ζ	A&G	109-1	33
Song Yingxing	IAU	LAC23	9
Sosigenes	IAU	LAC60	35
Sosigenes A	IAU	LAC60	35
Sosigenes B	IAU	LAC60	35
Sosigenes BA	AIC	60D	35
Sosigenes C	IAU	LAC60	35
Sosigenes CA	AIC	60D	35
Sosigenes CB	AIC	60D	35
South	IAU	LAC11	2
South A	IAU	LAC10	2
South B	IAU	LAC11	2
South C	IAU	LAC11	2
South Crater	CAF	B4	Detail 2
South D	IAU	LAC11	2
South E	IAU	LAC10	1, 2
South F	IAU	LAC10	1, 2
South G	IAU	LAC10	1, 2
South H	IAU	LAC11	2
South K	IAU	LAC10	2
South M	IAU	LAC10	1, 2
South Massif	IAU	LAC43	Detail 6
Spirit	CAF	B6	Detail 6
Spitzbergen A	IAU	LAC25	11
Spitzbergen C	IAU	LAC25	11
Spitzbergen D	IAU	LAC25	11
Spitzbergen α	A&G	115-1	12
Spitzbergen β	A&G	115-1	12
Spitzbergen γ	A&G	115-1	12
Spitzbergen δ	A&G	115-1	12
Spitzbergen ϵ	A&G	115-2	12
Spitzbergen κ	A&G	115-1	12
Spitzbergen μ	A&G	115-2	12
Spur	IAU	LAC41	Detail 2
Spurr	IAU	LAC41	22
Sputnik	CAF	B6	Detail 6

Feature	Source	Reference	Map(s)
St. George	IAU	LAC41	22, Detail 2
St. Teresa	CAF	B1	Detail 8
Stadius	IAU	LAC58	32
Stadius A	IAU	LAC58	32
Stadius B	IAU	LAC58	32
Stadius C	IAU	LAC58	32
Stadius D	IAU	LAC58	31, 32
Stadius E	IAU	LAC58	31, 32
Stadius F	IAU	LAC58	31, 32
Stadius G	IAU	LAC58	32
Stadius H	IAU	LAC58	32
Stadius J	IAU	LAC58	20, 21, 31, 32
Stadius K	IAU	LAC58	32
Stadius L	IAU	LAC58	32
Stadius M	IAU	LAC58	20, 31
Stadius N	IAU	LAC58	31, 32
Stadius P	IAU	LAC58	32
Stadius Q	IAU	LAC58	32
Stadius R	IAU	LAC58	32
Stadius S	IAU	LAC58	31, 32
Stadius T	IAU	LAC58	31, 32
Stadius U	IAU	LAC58	20, 21, 31, 32
Stadius W	IAU	LAC58	20, 21, 31, 32
Statio Tranquillitatis	IAU	LAC60	Detail 7
Stella	IAU	LAC42	25
Steno-Apollo	IAU	LAC43	Detail 6
Stewart	IAU	LAC62	38
Strabo	IAU	LAC14	6
Strabo B	IAU	LAC5	6
Strabo C	IAU	LAC5	6
Strabo L	IAU	LAC4	6
Strabo N	IAU	LAC5	6
Struve	IAU	LAC37	17
Struve B	IAU	LAC37	17
Struve C	IAU	LAC37	17
Struve D	IAU	LAC37	17
Struve F	IAU	LAC37	17
Struve G	IAU	LAC37	17
Struve H	IAU	LAC37	17
Struve K	IAU	LAC37	17
Struve L	IAU	LAC37	17
Struve M	IAU	LAC37	17
Suess	IAU	LAC57	29
Suess B	IAU	LAC57	29
Suess D	IAU	LAC57	29
Suess F	IAU	LAC57	29
Suess FA	A&G	144-1	29
Suess FB	A&G	144-1	29
Suess G	IAU	LAC57	29
Suess H	IAU	LAC57	29
Suess J	IAU	LAC57	29
Suess K	IAU	LAC56	29
Suess L	IAU	LAC56	29
Sulpicius Gallus	IAU	LAC42	23
Sulpicius Gallus A	IAU	LAC41	23

Feature	Source	Reference	Map(s)
Sulpicius Gallus B	IAU	LAC42	23
Sulpicius Gallus BA	A&G	97-3	23
Sulpicius Gallus BB	A&G	90-2	23
Sulpicius Gallus G	IAU	LAC41	23
Sulpicius Gallus H	IAU	LAC41	22, 23
Sulpicius Gallus M	IAU	LAC41	23
Sulpicius Gallus α	NLC	LAC42	23
Swift	IAU	LAC44	26
SWP	CAF	B6	Detail 6
Sylvester	IAU	LAC1	3
Sylvester N	IAU	LAC1	3
T. Mayer	IAU	LAC58	19
T. Mayer A	IAU	LAC58	19, 20
T. Mayer B	IAU	LAC57	19
T. Mayer C	IAU	LAC58	31
T. Mayer D	IAU	LAC58	30, 31
T. Mayer E	IAU	LAC40	20
T. Mayer F	IAU	LAC58	30
T. Mayer G	IAU	LAC40	19, 20
T. Mayer H	IAU	LAC58	31
T. Mayer J	A&G	126-2	20, 31
T. Mayer K	IAU	LAC40	19, 20
T. Mayer L	IAU	LAC58	31
T. Mayer M	IAU	LAC58	20
T. Mayer N	IAU	LAC58	31
T. Mayer P	IAU	LAC58	19, 30
T. Mayer R	IAU	LAC58	31
T. Mayer S	IAU	LAC58	30
T. Mayer W	IAU	LAC39	19
T. Mayer Z	IAU	LAC58	20, 31
T. Mayer α	A&G	133-2	19, 30
T. Mayer β	A&G	138-2	19
T. Mayer δ	A&G	133-2	19, 30
T. Mayer ζ	A&G	133-2	30
T. Mayer η	A&G	133-2	19, 20, 30, 31
T. Mayer κ	A&G	138-2	30
T. Mayer λ	A&G	133-2	19
T. Mayer ν	A&G	133-2	30
T. Mayer ξ	A&G	133-2	19
T. Mayer π	A&G	133-2	19, 30
T. Mayer ρ	A&G	138-2	19
T. Mayer σ	A&G	133-2	19
T. Mayer ω	A&G	133-2	19, 20, 31
Tacchini	IAU	LAC63	38
Tacquet	IAU	LAC42	24
Tacquet B	IAU	LAC60	24
Tacquet BA	A&G	85-2	24
Tacquet BB	A&G	90-2	24
Tacquet C	IAU	LAC60	35
Tai Wei	IAU	LAC24	11
Taizo	IAU	LAC41	22
Taruntius	IAU	LAC61	37
Taruntius B	IAU	LAC61	37
Taruntius CA	A&G	61-1	37
Taruntius CB	A&G	61-1	37

Feature	Source	Reference	Map(s)
Taruntius EB	A&G	66-1	37
[illegible]	[illegible]	[illegible]	[illegible]
Taruntius ED	AIC	61D	36
Taruntius F	IAU	LAC61	37
Taruntius H	IAU	LAC61	37
Taruntius K	IAU	LAC62	37
Taruntius L	IAU	LAC61	37
Taruntius MA	A&G	66-1	37
Taruntius MB	A&G	66-1	37
Taruntius O	IAU	LAC62	37, 38
Taruntius P	IAU	LAC62	37
Taruntius R	IAU	LAC61	37
Taruntius S	IAU	LAC61	37
Taruntius T	IAU	LAC61	37
Taruntius T*	IAU	LAC61	37
Taruntius TA	A&G	61-1	37
Taruntius U	IAU	LAC62	37
Taruntius V	IAU	LAC61	37
Taruntius VB	A&G	61-1	37
Taruntius W	IAU	LAC61	37
Taruntius WA	A&G	61-1	37
Taruntius WB	A&G	61-1	37
Taruntius X	IAU	LAC62	37
Taruntius Z	IAU	LAC61	37
Taruntius η	A&G	66-1	37
Taruntius θ	AIC	61D	36
Taruntius ι	AIC	61D	36
Taruntius κ	AIC	61D	36
Tassaday	CAF	B6	Detail 6
Tebbutt	IAU	LAC62	37
Tecumseh	CAF	B4	Detail 2
Tempel	IAU	LAC60	34
Tempel A	AIC	60D	34
Tempel AB	AIC	60D	34
Teneriffe α	A&G	127-3	11
Teneriffe γ	A&G	127-3	11
Teneriffe δ	A&G	127-3	11
Teneriffe ϵ	A&G	127-3	11
Teneriffe κ	A&G	127-3	11
Teneriffe ω	A&G	127-3	11
Terrace	IAU	LAC41	Detail 2
Thales	IAU	LAC14	6
Thales A	IAU	LAC14	6
Thales E	IAU	LAC14	6
Thales F	IAU	LAC14	6
Thales G	IAU	LAC14	6
Thales H	IAU	LAC14	6
Thales W	IAU	LAC13	6
The Cape	CAF	B1	Detail 8
The Gashes	CAF	B1	Detail 8
The Triangle	CAF	B1	Detail 8
The Trio	CAF	B1	Detail 8
The 'Z'	CAF	B1	Detail 8
Theaetetus	IAU	LAC25	12
Theiler	IAU	LAC63	38
Theon Senior	IAU	LAC78	34
[illegible]	[illegible]	[illegible]	[illegible]
Theon Senior B	IAU	LAC60	34
Theon Senior BA	AIC	60D	34
Theophrastus	IAU	LAC43	25
Thud Ridge	CAF	B1	Detail 8
Tian Shi	IAU	LAC24	11
Timaeus	IAU	LAC12	4
Timocharis	IAU	LAC40	21
Timocharis AA	A&G	122-1	21
Timocharis B	IAU	LAC40	21
Timocharis C	IAU	LAC40	21
Timocharis D	IAU	LAC40	21
Timocharis E	IAU	LAC40	20, 21
Timocharis H	IAU	LAC40	20, 21
Tisserand	IAU	LAC43	26
Tisserand A	IAU	LAC43	26
Tisserand B	IAU	LAC44	26
Tisserand D	IAU	LAC43	26
Tisserand K	IAU	LAC44	26
Tortilla Flat	IAU	LAC43	Detail 6
Toscanelli	IAU	LAC39	18
Towers	CAF	B6	Detail 6
Townley	IAU	LAC62	38
Tralles	IAU	LAC44	26
Tralles A	IAU	LAC43	26
Tralles B	IAU	LAC44	26
Tralles C	IAU	LAC43	26
Trident	IAU	LAC43	Detail 6
Triesnecker	IAU	LAC59	33
Triesnecker D	IAU	LAC59	34
Triesnecker E	IAU	LAC59	33
Triesnecker F	IAU	LAC59	33
Triesnecker G	IAU	LAC59	33, 34
Triesnecker H	IAU	LAC59	33
Triesnecker J	IAU	LAC59	33
Trophy Point	CAF	B4	Detail 2
Trouvelot	IAU	LAC12	4
Trouvelot D	A&G	116-1	4
Trouvelot G	IAU	LAC25	12
Trouvelot H	IAU	LAC12	4
Trouvelot η	A&G	110-2	12
Trouvelot ξ	A&G	110-2	12
Turner A	IAU	LAC76	32
Turner B	IAU	LAC76	32
Turner Q	IAU	LAC76	32
Turner λ	A&G	114-1	32
Turner μ	A&G	114-1	32
Twin Craters Ridge	CAF	B1	Detail 8
U.S. 1	CAF	B1	Detail 8
Ukert	IAU	LAC59	33
Ukert A	IAU	LAC59	33
Ukert B	IAU	LAC59	33
Ukert D	A&G	102-1	33
Ukert E	IAU	LAC59	33

Feature	Source	Reference	Map(s)
Ukert J	IAU	LAC59	33
Ukert K	IAU	LAC59	33
Ukert M	IAU	LAC59	33
Ukert N	IAU	LAC59	33
Ukert P	IAU	LAC59	33
Ukert R	IAU	LAC59	33
Ukert V	IAU	LAC59	33
Ukert W	IAU	LAC59	33
Ukert X	IAU	LAC59	33
Ukert Y	IAU	LAC59	33
Ulugh Beigh	IAU	LAC22	8
Ulugh Beigh A	IAU	LAC22	8
Ulugh Beigh B	IAU	LAC22	8
Ulugh Beigh C	IAU	LAC37	8
Ulugh Beigh D	IAU	LAC37	8
Ulugh Beigh M	IAU	LAC22	8
Urey	IAU	LAC45	27
Uttermost West	CAF	B4	Detail 2
Väisälä	IAU	LAC39	18
Valera	IAU	LAC24	Detail 1
Vallis Alpes	IAU	LAC12	4, 12
Vallis Christel	IAU	LAC42	Detail 4
Vallis Krishna	IAU	LAC42	Detail 4
Vallis Schröteri	IAU	LAC39	18
Van Albada	IAU	LAC62	38
Van Biesbroeck	IAU	LAC38	18, 19
Van Serg	IAU	LAC43	Detail 6
Vasco da Gama	IAU	LAC55	17, 28
Vasco da Gama A	IAU	LAC55	28
Vasco da Gama B	IAU	LAC55	17
Vasco da Gama C	IAU	LAC55	28
Vasco da Gama F	IAU	LAC55	17, 28
Vasco da Gama P	IAU	LAC55	28
Vasco da Gama S	IAU	LAC55	28
Vasco da Gama T	IAU	LAC55	28
Vasya	IAU	LAC24	Detail 1
Vera	IAU	LAC39	19
Verne	IAU	LAC40	20
Very	IAU	LAC42	24
Victory	IAU	LAC43	Detail 6
Virchow	IAU	LAC63	38
Vitruvius	IAU	LAC43	25
Vitruvius B	IAU	LAC43	25
Vitruvius C	A&G	78-2	25
Vitruvius G	IAU	LAC61	25, 36
Vitruvius H	IAU	LAC43	25
Vitruvius K	A&G	78-2	25
Vitruvius L	IAU	LAC43	25
Vitruvius M	IAU	LAC43	25
Vitruvius T	IAU	LAC43	25
Vitya	IAU	LAC24	Detail 1
Volta	IAU	LAC21	1
Volta B	IAU	LAC21	1
Volta D	IAU	LAC21	1
Von Braun	IAU	LAC22	8

Feature	Source	Reference	Map(s)
Voskresenskiy K	IAU	LAC37	17
W. Bond	IAU	LAC3	4
W. Bond B	IAU	LAC3	4
W. Bond C	IAU	LAC3	4
W. Bond D	IAU	LAC12	4
W. Bond E	IAU	LAC12	4
W. Bond F	IAU	LAC3	4
W. Bond G	IAU	LAC12	4
Wagner	CAF	B6	Detail 6
Wagon Road	CAF	B1	Detail 8
Walden	CAF	B6	Detail 6
Wallace	IAU	LAC41	21
Wallace A	IAU	LAC41	21, 22
Wallace C	IAU	LAC41	21
Wallace D	IAU	LAC41	21, 22
Wallace H	IAU	LAC41	21
Wallace K	IAU	LAC41	21
Wallace T	IAU	LAC41	22
Wallach	IAU	LAC61	36
Walter	IAU	LAC39	19
Wash Basin	CAF	B1	Detail 8
Watts	IAU	LAC61	37
Weatherford	CAF	B1	Detail 8
Webb	IAU	LAC80	38
Webb B	IAU	LAC80	38
Webb C	IAU	LAC62	38
Webb E	IAU	LAC62	38
Webb F	IAU	LAC62	38
Webb G	IAU	LAC62	38
Webb J	IAU	LAC80	38
Webb K	IAU	LAC80	38
Webb L	IAU	LAC62	38
Webb M	IAU	LAC80	38
Webb N	IAU	LAC80	38
Webb P	IAU	LAC62	38
Webb Q	IAU	LAC80	38
Webb U	IAU	LAC62	38
Webb W	IAU	LAC62	38
Webb X	IAU	LAC62	38
Wegener	IAU	B6	Detail 6
Weierstrass	IAU	LAC81	38
Wessex Cleft	IAU	LAC43	Detail 6
West	IAU	LAC60	Detail 7
Whewell	IAU	LAC60	34
Whewell A	IAU	LAC60	34
Whewell B	IAU	LAC60	34
Whewell BA	AIC	60D	34
Wilbur	CAF	B4	Detail 2
Wildt	IAU	LAC63	38
Williams	IAU	LAC27	14, 15
Williams F	IAU	LAC27	14
Williams M	IAU	LAC27	15
Williams N	IAU	LAC27	14
Williams R	IAU	LAC27	14, 15
Window	CAF	B4	Detail 2

Feature	Source	Reference	Map(s)
Wolff A	IAU	LAC59	21
Wolff B	IAU	LAC41	21
Wollaston	[illegible]	[illegible]	8, 9
Wollaston D	IAU	LAC23	8, 9
Wollaston N	IAU	LAC39	18
Wollaston P	IAU	LAC39	18
Wollaston R	IAU	LAC38	18
Wollaston U	IAU	LAC38	8
Wollaston V	IAU	LAC38	8
Wollaston α	A&G	151-1	9, 18, 19
Wollaston γ	A&G	158-2	9
Worm Rille	CAF	B1	Detail 8
Xenophanes	IAU	LAC21	1
Xenophanes A	IAU	LAC21	1
Xenophanes B	IAU	LAC21	1
Xenophanes C	IAU	LAC10	1
Xenophanes D	IAU	LAC10	1
Xenophanes E	IAU	LAC21	1
Xenophanes F	IAU	LAC10	1
Xenophanes G	IAU	LAC10	1
Xenophanes K	IAU	LAC21	1
Xenophanes L	IAU	LAC10	1
Xenophanes M	IAU	LAC10	1
Xu Guangqi	IAU	LAC23	9
Yangel'	IAU	LAC41	22
Yerkes	IAU	LAC62	26, 37
Yerkes E	IAU	LAC62	26
Yerkes V	IAU	LAC62	26
Yoshi	IAU	LAC42	Detail 4
Zähringer	IAU	LAC61	36, 37
Zeno	IAU	LAC28	16
Zeno A	IAU	LAC28	16
Zeno B	IAU	LAC28	16
Zeno D	IAU	LAC28	16
Zeno E	IAU	LAC28	16
Zeno F	IAU	LAC28	16
Zeno G	IAU	LAC28	16
Zeno H	IAU	LAC28	16
Zeno J	IAU	LAC28	16
Zeno K	IAU	LAC28	16
Zeno P	IAU	LAC28	16
Zeno U	IAU	LAC28	16
Zeno V	IAU	LAC28	16
Zeno W	IAU	LAC28	16
Zeno X	IAU	LAC28	16
Zi Wei	IAU	LAC24	11
Zinner	IAU	LAC38	18

References used for the creation of this book

Gary L. Gutschewski, Danny C. Kinsler, Ewen Whitaker
Atlas and Gazetteer of the Near Side of the Moon
Scientific and Technical Information office National Aeronautics and Space Administration, Washington D.C., USA, 1971

Mary A. Blagg and S.A. Saunder
Collated List of Lunar Formations
Messrs Neill & Co., Ltd., Edinburgh, UK, 1913

R. V. Wagner, D. M. Nelson, J.B. Plescia, M. S. Robinson, E. J. Speyerer, E. Mazarico,
Coordinates of Anthropogenic Features on the Moon
Elsevier, Tempe, AZ, USA, 2016

John Moore
Craters of the Near Side Moon
2014

John Moore
Features of the Near Side Moon
2017

Gazetteer of Planetary Nomenclature Homepage
http://planetarynames.wr.usgs.gov/

LPI Resources
https://www.lpi.usra.edu/resources/mapcatalog/

LROC Lunar Quickmap
https://quickmap.lroc.asu.edu/

LROC Website
https://www.lroc.asu.edu/

Robert A. Garfinkle
Luna Cognita – A Comprehensive Observer's Handbook of the Known Moon.
Volume 1, 2 & 3
Springer, New York, USA, 2020

Ewen A. Whitaker
Mapping and Naming the Moon
Cambridge University Press, Cambridge, UK, 1999

Antonín Rükl
Mondatlas
Aventinum Nakladatelství, Prag, Czech Republic, 2012

Mary A. Blagg, K. Müller
Named Lunar Formations
Percy Lund, Humphries & Co. Ltd., London, UK, 1935

Leif E. Andersson, Ewen A. Whitaker
"NASA Catalogue of Lunar Nomenclature" (RP-1097)
University of Arizona, Tuscon Arizona, USA, 1982

D. W. G. Arthur, Alice P. Agnieray, Ruth A. Horvath, C. A. Wood and C. R. Chapman
Nos. 30, 40, 50 & 70. The System of Lunar Craters, Quadrant 1, 2, 3 & 4
1963-1966

Made in United States
Troutdale, OR
01/12/2024

16919534R00081